SpringerBriefs in Electrical and Computer Engineering

For further volumes:
http://www.springer.com/series/10059

Jesse M. Lingeman · Dennis Shasha

Network Inference in Molecular Biology

A Hands-on Framework

 Springer

Jesse M. Lingeman
New York University
4 Washington Place
New York, NY 10003
USA

Dennis Shasha
New York University
251 Mercer Street
New York, NY 10012
USA

ISSN 2191-8112   ISSN 2191-8120   (electronic)
ISBN 978-1-4614-3112-1   ISBN 978-1-4614-3113-8   (eBook)
DOI 10.1007/978-1-4614-3113-8
Springer New York Heidelberg Dordrecht London

Library of Congress Control Number: 2012939634

Printed on acid-free paper

Springer is part of Springer Science+Business Media (www.springer.com)

Acknowledgements

We would like to thank Alex Greenfield, Rich Bonneau, and Alex Rubinsteyn for their comments both technical and stylistic. We would also like to thank Juliana Freire for her help with workflow concepts and the Vistrails software.

This work has been partly supported by the US National Institutes of Health grant 2R01GM032877-25A1 and the US National Science Foundation grants MCB-0929339, IOS-0922738, and IIS - 1050388. That support is greatly appreciated.

Contents

Chapter 1
Overview of Network Inference

1.1 Inferring Causality

Inferring a causal link is useful in many applications, from medicine to economics to engineering. If some A can cause some B to take on a high value (where A could be a gene in our context, a market intervention in economics, or a design element in engineering), then preventing B from taking such a value can be done by removing some B, by removing some A or by interfering with the link from A to B. Conversely, making B achieve a higher value can be done by adding more B, adding more A, or enhancing the efficiency of the link from A to B.

Approaches to finding such causal links may entail performing:

- experiments under multiple different conditions to detect associations between A and other elements. Such associations are bi-directional but may acquire directionality if it is known that, for example, A is an element that can change other elements (in the genomic context A could be a "transcription factor") and B is not.
- experiments that increase the quantity of some A to see which other elements are either enhanced (quantity increases) or repressed (quantity decreases).
- experiments that decrease the quantity of A or even knock it out (A goes to 0) may also reveal something about the influence of A.
- experiments over a closely spaced time course to enable inferences of the form "the state of A at time t may influence B at time t+1."

This book is about inference of causality in genomics and more generally molecular biology, but its techniques apply to any setting in which elements may singly or collectively affect others. The set of genomic algorithms provides a variety of analytic methods for each experimental setup described above. If you come from another field, you may find some of the concepts strange, so we review them here. Don't worry. There aren't many.

J. M. Lingeman and D. Shasha, *Network Inference in Molecular Biology,*
SpringerBriefs in Electrical and Computer Engineering,
DOI: 10.1007/978-1-4614-3113-8_1, © The Author(s) 2012

1.1.1 Basics of Genomics

In general, genomics is focussed on all the genetic material (i.e. all the DNA) in a species. Genetic material concerns both genes and the DNA between genes. The main question we will address in this book is "which genes influence which other genes and by how much?" Here is the main biological terminology we will use.

Nucleotide (also known as base)   One of four molecules, labeled as A, C, T, or G. This is the smallest distinguishable element of DNA.

Gene   A sequence of DNA which can form RNA. A gene is preceded by a *promoter* which is also DNA. DNA has a directionality, so the notion of preceding is well-defined.

Transcription   If appropriate proteins (called *transcription factors*) bind to the promoter, then a copy of the DNA is made into RNA. (This copy is really a close copy because RNA nucleotides are slightly different from DNA ones, but mathematically this is just a copy, because they correspond 1 to 1.) This is done without destroying the DNA. The copying of DNA into RNA is called *transcription*.

Expression of a gene g   The amount of RNA of gene g present in a solution.

Translation   The process by which RNA is made into a protein. Not all RNA is made into a protein, so knowing the levels of RNA does not imply knowledge of the associated protein. As of this writing, most quantitative information is a measure of RNA expression. The algorithms wouldn't change if we had knowledge of the proteins (especially the transcription factors), but the results would be better.

Genome   The genes and their associated promoters.

Causality   One gene A can influence another gene B "transcriptionally" as follows: if gene A is transcribed to RNA and then translated to a protein, then the resulting protein may bind to the promoter of B and either help cause B to transcribe or inhibit it from transcribing. Because there are tens of thousand of genes in many organisms, the possible causal relationships are in the hundreds of millions. Fortunately, many genes are tightly correlated to one another and are often partly redundant. That fact allows our analytical programs to group genes into clusters that can be treated one as one unit.

Fig. 1.1 Proteins are born in a two step process. Genes are "transcribed" to RNA which is "translated" to proteins. Experimental techniques currently permit the measurement of the expression levels of most of the RNA for a species. The measurement of protein levels should make the analysis better.

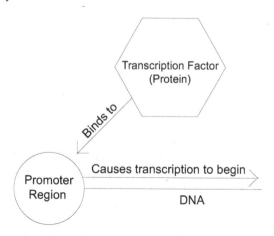

Fig. 1.2 Certain proteins called transcription factors bind to promoters. Each transcription factor can either induce or repress transcription. If a good combination of transcription factors bind to the promoter of a gene, then that gene may be transcribed to RNA.

1.1.2 Observation, Intervention, and Inference

Behind every causal conclusion, there must lie some causal assumption. [25] [18] In our case, one causal assumption is that the state of the system at one time point determines the state of the system at the next time point if the time points are close enough (within a few minutes). Another causal assumption is that when a system is in a steady state and then an external intervention (e.g. a gene knockout) either increases or decreases the RNA of a gene, then any changes from steady state are due to that intervention.

In very special circumstances, one can infer that some set of genes Y will take on some values V_y (say, of expression) based on forcing the expressions of genes X to V_x. The circumstances are that one observes Y to take on V_y when X takes on V_x (but without forcing X to do so) AND whatever influences Y consists of X and elements that are independent of the values of the X genes. Without an external intervention, this assumption nearly never holds, because biological systems are full of feedback loops, so anything that influences Y will also likely influence X. That is why the most informative experiments incorporate an intervention, whereas purely observational experiments (e.g. experiments based on steady conditions) are never conclusive. In this book, we use observational experiments to assign higher initial weights to some edges than to others, but then the iterative learning methods we discuss may change those weights based on evidence with known causal directions (such as time series or knockouts).

1.1.3 The Data

A basic gene expression dataset is a matrix where columns represent different conditions (or experiments) and rows represent different genes. The value of each cell in the matrix is the expression value of a given gene in a given condition. A value of 0 means that there is no detectable gene expression. A higher value represents a higher gene expression. Whether a value is "high" or "low" is relative to the other values in that dataset. The absolute values of these numbers are determined by the physical device and preprocessing methods used to gather the data. What is important are the relative values. Typically, the values are normalized by column to be between 0 and 1.

There are four different basic types of data: steady-state, time-series, multifactorial, and wildtype. Data are considered steady-state if the values are not changing or changing very slowly. A perturbation can be a chemical (e.g., as adding a nutrient, like nitrogen, to a plant) or genetic (e.g., removing or repressing a specific gene) in nature. Steady-state values as used in this experiment are often the difference between the steady state values after a perturbation and the difference of some reference condition prior to the perturbation.

Time-series data are similar to steady-state data in that a genetic or chemical perturbation is applied to a set of genes, but then readings are taken while the expression values of each gene are still changing. This gives us extra information about how the expression values of the genes actually get from one steady-state value to another.

Multifactorial data are simply steady-state datasets where more than one perturbation has occurred. For example, two genes could be knocked out of the dataset instead of just one, or there could be both a genetic and chemical perturbation. Wild-type data are expression values from the most common phenotype of a given organism.

1.1.4 Data Sources

To generate data for the algorithms, we use a data simulator called GeneNetWeaver [22] [27] as well as real data. GeneNetWeaver was developed in order to generate simulated datasets for the Dialogue for Reverse Engineering Assessments and Methods (DREAM) competition, and has been released as free software. Given a network graph, GeneNetWeaver builds a dynamical model, generates expression values for each gene, and then adds noise similar to what is found in experimental data. We have used simulated rather than real data to be able to test the algorithms against a known ground truth (in the form of a known causality network).

Real experimental data can be found in many public online databases, such as Arabidopsis.org or GenExpDB.

Because we are testing these algorithms against simulated data, the conclusions about the quality of the algorithms should be taken with a certain skepticism. If you can build your own benchmark to reflect the kind of data your application produces, then you can easily retry these algorithms to see which one will likely work best for you. [1]

1.1.5 Basic Data Metrics

In order to compare performance of algorithms and pipelines we use some simple statistics. All of these are calculated by comparing each edge of an inferred network to a gold standard (in our case, usually the underlying "true" network generated by the DREAM simulator).

True Positive (TP)   The existence of an edge between two genes is predicted and that is correct based on the gold standard.

False Positive (FP)   The existence of an edge between two genes is predicted but that is incorrect based on the gold standard.

True Negative (TN)   The non-existence of an edge between two genes is predicted and that is correct, i.e. there is no edge between those genes in the gold standard network.

False Negative (FN)   The non-existence of an edge between two genes is predicted but that is incorrect, i.e. there is an edge between those genes in the gold standard network.

Precision   The ratio of true positives to the number of true positives plus the number of false positives. Indicates how often a positive prediction was correct out of the total number of positive predictions. Sometimes called the positive predictive value (PPV).

$$\text{precision} = \frac{\text{true positives}}{\text{true positives} + \text{false positives}} \tag{1.1}$$

Accuracy   How many positive or negatives inferences were correct out of all positive and negative predictions.

$$\text{accuracy} = \frac{\text{true positives} + \text{true negatives}}{\text{true positives} + \text{false positives} + \text{true negatives} + \text{false negatives}} \tag{1.2}$$

Recall   The ratio of true positives to all positives, i.e., the number of correctly inferred positive edges vs. the number of positive edges in the gold standard network.

$$\text{recall} = \frac{\text{true positives}}{\text{true positives} + \text{false negatives}} \tag{1.3}$$

Precision-Recall Curve   The precision-recall curve is a way of visualizing the ratio of correct to incorrect positive guesses vs. the ratio of correct positive edges to

---

[1] DREAM is a yearly competition that aims to challenge researchers to develop better algorithms. Each year a more difficult dataset is released, and the challenge is to develop and algorithm that tries to reconstruct a gene regulatory network from the given data. The algorithms are then sent to the DREAM organizers and the output of each algorithm is evaluated and ranked.

the total number of positive edges in the gold standard. To do this, a ranked list of edges is needed. Edges are ranked in descending order of the inferred likelihood that they are positive, and then they are plotted based on that order. Specifically, the precision and the recall of the first k edges in descending order is the kth point in the precision-recall curve. This gives a picture of what happens as we move from the most likely edges to the least.

The precision-recall curve tells us at what point the algorithm becomes unreliable. For example, in Figure 1.3, we can see that the first several inferred edges are correct (precision is staying at 1.0 as the recall goes up), and then drops it once some of the inferred edges turn out to be false positive. For some algorithms, we may see that the highest ranked edges are very accurate, but the quality of the algorithmic guesses drops off rapidly. A precision-recall graph would show this clearly, where a ROC curve (the next definition) may not show this behavior at all.

In practice, there will be a threshold in the ranking that will determine which edges to keep. That threshold can be determined by looking at the ROC curve on similar data for which the answers are known. Most model species have at least some such gold standard data. At other times, experimentalists may simply decide to do tests in the order of the ranking of the edges, thus creating their own gold standard data.

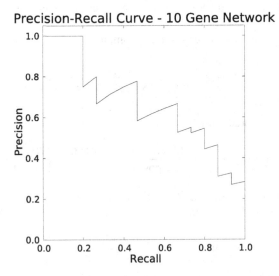

Fig. 1.3 An example precision-recall curve. The first few inferred edges are correctly inferred, so precision stays at 1. After a few edges, the precision starts to drop, slightly increasing with each correctly inferred edge and slightly decreasing with incorrect edges.

Receiver-Operating Characteristic (ROC) Curve   The ROC curve is a way of visualizing overall performance by comparing the true positive rate vs. the false positive rate (Figure 1.4). The true positive rate is the same as recall. The false positive rate is equal to the number of false positives over the sum of false positives and true negatives. To plot the curve, edges are ranked in descending order of the inferred likelihood that they are positive and then plotted. The higher the "arc" of the curve (the more convex upwards), the better the performance, as ideally the true positive rate rises quickly while the false positive rate stays at or near 0. The area under the receiver-operating characteristic curve (AUROC) will be used as a single number measure for the performance of an algorithm. An AUROC of 0.5 means that the algorithm has a 0.5 chance of being correct with each positive guess, in which case the curve will fall along the 45 degree diagonal of the graph.

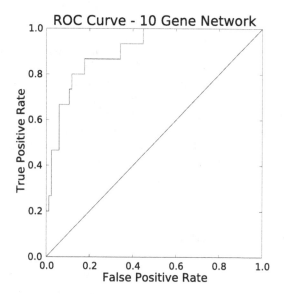

Fig. 1.4 An example ROC curve. This graph has a very high "arc": the true positive rate is rising very quickly compared to the false positive rate.

p-value If an inference method yields some result (e.g. 60% of the positively predicted edges are correct), an important question to ask is whether this could have happened just by random chance. If so, then the inference method is not so useful. If not, then the inference method may be very valuable. The *p*-value is the probability that a result could have been obtained just by random chance. How this is calculated depends on the application.

z-score A collection of values has a mean m and standard deviation s. The z-score of a particular value v is the number of standard deviations above or below the mean m, i.e. $\text{zscore}(v) = (v - m)/s$.

1.1.6 How to Read This Book

We have separated the chapters into three steps, followed by ways to wrap the steps into one of several possible pipelines. Within each chapter describing a step, we present alternative ways of performing that step and give an indication of what works best under which circumstances.

* Chapter 1 gives different techniques for clustering genes and expressions. Clustering effectively creates "supernodes" out of many genes, thus both simplifying the network inference problem and the interpretation of the result.

* Chapter 2 talks about using steady state data resulting from knock-outs, knock-downs, over-expressors, or just particular conditions. Such data can suggest causal or at least correlative relationships among pairs of genes.

* Chapter 3 discusses the analysis of time course data. The assumption driving that analysis is that events at a current time point depend only on the data in previous time points.

* Chapter 4 describes various ways in which the algorithms of the first three chapters can be combined into a full analytical pipeline.

Each algorithm includes an experimental evaluation on data from the simulated biological world of DREAM. Some algorithms do better than others on those examples. We indicate the quality of each algorithm on the DREAM simulation data at the beginning of each section. As we said above, an inference algorithm may work better or worse on other data. This does not necessarily reflect their use on problems that you might care about, but to give you an idea of their performance, each inference algorithm is ranked from Fair to Best.

1.1.7 How to Use the Software

We have included a software package to be downloaded with this text, available at http://cs.nyu.edu/shasha/netinf. The software may be obtained in the form of a virtual machine with all of the packages pre-installed, or a standalone package that requires some installation. The virtual machine can be launched in the free, open-source software VirtualBox (http://www.virtualbox.org).

To launch the virtual machine in VirtualBox, first unzip the zip file containing the virtual machine. This will give you a folder called NetworkInference. Open VirtualBox and select "Import Appliance" from the "File" menu, and choose the NetworkInference.ovf file in the unzipped virtual machine directory. After some loading, VirtualBox will import the virtual machine. Once the virtual machine appears in VirtualBox's library, click "Start".

Once the virtual machine boots, you will see a Linux desktop. Open a terminal and type cd NetworkInference to change to the network inference directory. This is the directory from which all scripts should be run.

This software wraps around the original implementations of each algorithm by the author, providing an easy to use interface with a common data format for all algorithms presented in this text. A script for each experiment performed in the book can be found in the folder "Examples". To invoke a script, launch the script from a command line with python. For example:

```
python Examples/MCZ/MCZ-dream4-10gene.py
```

In this case, MCZ is the algorithm name and the dataset is the 10 gene DREAM4 network. These scripts can serve as examples for creating new experiments. Please see the README files included with the software for more documentation.

Chapter 2
Step 1: Clustering Data

2.1 Introduction

2.1.1 Clustering

Clustering reduces the size of the data by replacing individual genes with artificial super-genes that can be treated as single nodes for the purposes of network inference. By clustering genes that work together as a preprocessing step, we can improve the accuracy of the resulting network by reducing variance due to noise on individual genes. The goal is to generate clusters while losing the minimum amount of information in the dataset (and perhaps even make certain relationships stronger!). For example, if there are two genes that both behave in exactly the same way across the experimental conditions of interest, then little to no information is lost if you treat them as though they were a single "gene".

There are many different clustering algorithms available. The basic idea in all clustering algorithms is to group data that are alike together based on some measure of similarity. A classic example is the K-Means algorithm. K-Means takes a parameter k that specifies the number of clusters to generate. In the random start method, K "centroids" are chosen at random (e.g. these may be the expression profiles of k genes). Then (step 1) each expression profile is compared with all the centroids and is assigned to the most similar one. The expression profiles corresponding to a given centroid are called a cluster. Next (step 2), for each cluster a centroid that better represents the expression profiles of that cluster is chosen. Steps 1 and 2 are repeated until there is no further change in cluster membership. Then the whole procedure begins again with a new set of initial centroids. After trying something like 100 different sets of centroids, the algorithm starts with the best clustering found, where best means that each centroid is closest on the average to all the members. [1]

---

[1] Other schemes such as kmeans++ have also been proposed [1], but we like the random starting point approach for its sheer simplicity.

J. M. Lingeman and D. Shasha, *Network Inference in Molecular Biology*,
SpringerBriefs in Electrical and Computer Engineering,
DOI: 10.1007/978-1-4614-3113-8_2, © The Author(s) 2012

2.1.2 Biclustering

Biclustering is a method of clustering across both rows and columns, i.e. across gene expressions and conditions. Biclustering begins with a matrix where each column corresponds to an experiment and each row is a gene. Traditional clustering attempts to shuffle the rows into groups based on their expression values. However, this works poorly when grouping genes by expression values for different experiments, because genes may behave quite differently across experimental conditions, as we have mentioned above. Biclustering solves this by allowing the shuffling of both genes and experiments. This allows clusters to be selected based on a subset of experimental conditions rather than all of the experimental conditions in a given gene's row.

Genes may end up in more than one bicluster. Biologically, this makes sense, as a single gene may interact with different sets of genes in different conditions. For example, gene X may be part of cluster Y under condition A, but part of cluster Z under condition B.

2.2 cMonkey

2.2.1 What it Does

Finding the best biclustering (according to virtually any reasonable criterion) is NP-complete. cMonkey [26] combines biological intuition with heuristics to come to high quality biclusters in all cases we have tested.

2.2.2 The Data

The data used can be broken up into three types: expression data, upstream DNA sequence data, and (where available) experimentally verified network connections. The expression data are steady-state data, though time-series data can be used as well.

2.2.3 The Strategy

The algorithm starts by seeding biclusters with genes that are co-expressed across most if not all conditions (this is the basic strategy, though there are variants). After that, a gene or condition can be associated with a bicluster based on a score that is formulated as a probability of association with the current members of that bicluster. A gene can be associated with many biclusters depending on whether its score exceeds a certain threshold (using several different criteria). As the membership of biclusters changes, a particular gene may move in or out of a particular bicluster.

The criteria to determine whether a gene and/or experiment should belong with a bicluster are expression similarity, promoter similarity, and known network connections. We first show how to calculate the score that determines whether a gene or experimental condition should be associated with a bicluster based on expression (2.2.3.1). Second we describe the calculation of a score based on DNA motifs (small binding portions of a promoter) to determine whether the promoter of a gene warrants that gene's association with a bicluster (2.2.3.2). Third we describe how to use known network edges (e.g. metabolic edges) to determine whether a gene should be associated with a bicluster (2.2.3.3). Finally, we show how all these different criteria can be combined using a heuristic technique called simulated annealing into a single regression score (2.2.3.4 and Equation (2.4)).

The basic idea is that at each step of the algorithm, we calculate the probability that each gene g and each experiment e belongs in bicluster k, given the genes and experiments already in k. Based on that probability, the gene-experiment may be added to the bicluster. Biclusters are initialized by randomly choosing one of several methods (discussed in section 2.2.3.4). The probability of movement decreases over time (in the spirit of simulated annealing). The algorithm converges to a set of stable biclusters. The probability that a gene or condition should belong to a given bicluster is calculated for each of the types of data used at each iteration. These probabilities are then combined using a regression model, giving an overall probability that a given gene belongs to each bicluster.

We present the details of the algorithm below as four discrete steps. First, we calculate a likelihood score for each type of data separately. Then, we combine the scores into one value. Finally, we go over cMonkey's iterative procedure, putting all of the previous steps together. The net result is the likelihood a gene or condition/experiment should belong to a given bicluster.

2.2.3.1 Using the Expression Data

Expression data are used to create a likelihood that a given gene or experiment "belongs" to a given bicluster. We calculate the likelihood that x_{ij}, gene i's expression value in experiment j, is in a bicluster k with:

$$p(x_{ij}) = \frac{1}{\sqrt{2\pi\left(\sigma_j^2 + \varepsilon^2\right)}} \exp\left(-\frac{1}{2}\frac{\left(x_{ij} - \bar{x}_{jk}\right)^2 + \varepsilon^2}{\sigma_j^2 + \varepsilon^2}\right), \qquad (2.1)$$

where σ_j^2 is the variance of experiment j, ε is an error term, representing unknown error in the expression values, and $\bar{x}_{jk}$ is the mean expression level of experiment j over the genes in bicluster k. The variance over all of j is used instead of the variance of only the genes in bicluster k. This is done to help weed out experiments where there is not much variation between genes, i.e., experiments where genes are more likely to be correlated by random chance.

Once we have the likelihood that each measurement x_{ij} belongs to each bicluster k, we can calculate the likelihood that each gene and experiment belongs to a given bicluster. To calculate the likelihood that a gene i belongs to a bicluster, we take the product of the likelihoods that gene i is in bicluster k across all experiments j.

$$\text{prob gene } i \text{ belongs to bicluster } k = \prod_{j \in J_k} p(x_{ij}). \qquad (2.2)$$

Similarly, we calculate the likelihood that an experiment j belongs to a bicluster by taking the product of the likelihoods that experiment j is in bicluster k across all genes i.

$$\text{prob condition } i \text{ belongs to bicluster } k = \prod_{i \in I_k} p(x_{ij}). \qquad (2.3)$$

cMonkey then assigns a gene to a bicluster over the conditions of that bicluster if the probability of that gene belonging to a bicluster is greater than a randomly generated value between 0 and 1. Finally, we create a co-expression p-value for each gene i with respect to each bicluster k and for each experiment j with respect to each bicluster k, labeled r_{ik} and r_{ij}, respectively. These values are created by integrating over the normal distribution based on (2.1).

2.2.3.2 Using the Upstream DNA Sequence Data

Upstream DNA sequence data are used to identify whether or not a gene i shares DNA motifs with other genes in a bicluster k. A DNA motif is a sequence of DNA bases that are likely to be a target of transcription factors. So if two genes share the same motifs then they may have similar biological functions. The MEME algorithm [2] is used to identify these motifs, and the counterpart algorithm MAST [2] is used to calculate the p-value that a given sequence matches motifs found by MEME. MEME is used to create a set of the motifs found in the genes in each bicluster. MAST is then used to calculate a p-value s_{ik} of how likely it is that a gene i contains the motifs found in bicluster k. Intuitively, what we are checking is whether or not a gene shares similar biological functions (based on their DNA) with the rest of the genes in a given bicluster.

2.2.3.3 Using the Association Network Data

cMonkey also uses known association network data to help create biclusters. The network data can be obtained from the KEGG [16], Predictome [23] and Prolinks [3] databases. These datasets contain known network associations between genes. The basic idea here is to add genes to a cluster based on how many network associations that gene has in common with the genes in that cluster. The more associations a gene has in common with a cluster, the more likely it is that the gene belongs in that cluster. A p-value q_{ik}^n is calculated for each gene/cluster/network set, where n is each one of the different types of association networks (e.g., KEGG or Prolinks).

2.2.3.4 Putting it all together: The cMonkey Iterative Procedure

cMonkey starts by "seeding" each bicluster with an initial set of genes and experiments. Five different methods are used for seeding the initial biclusters:

1: A single random gene
2: Using co-expressed genes from another clustering method
3: Using semi-co-expressed genes (by correlation of expression values)
4: Using highly connected genes (from the association network)
5: Using genes with a common motif (from the upstream DNA sequence)

Many seeding strategies are used in order to introduce variance into the initialization of the algorithm. This helps keep the algorithm from getting caught in a local minimum early on.

cMonkey uses simulated annealing to select the biclusters. Simulated annealing is a probabilistic global optimization algorithm where a temperature parameter starts "hot" and "cools" over time. When the temperature is hot, genes and experiments are allowed to move much more freely between biclusters. This allows many combinations to be formed early on. As the temperature cools, moves between clusters become less and less likely, based on how "good" the move is. A gene that has a high affinity with a bicluster, as then constituted, has a higher probably of moving to it than one that doesn't. As the temperature cools, the biclusters "harden", until only the genes that are extremely good matches for a bicluster have any chance of actually moving to that bicluster. The maximum number of moves that can be made at each iteration is set by a parameter to be a small value, by default 5. Establishing a maximum is done to ensure that a bicluster cannot change too much in a single iteration.

In addition to the temperature parameter used in simulated annealing, the weights on the importance of each data type also change at each iteration. For example, Reiss et al. state that early in the procedure, DNA motifs are not very useful, as it is unlikely that any particular cluster has enough similar motifs to give a reliable signal. However, some of the association network data can be extremely informative early on. Thus, at the beginning of the algorithm, association network data are given a higher weight than the motif data. As the annealing continues and clusters become more and more numerous, using DNA motifs makes more sense, so the weight on DNA motifs is increased over time.

A constrained logistic regression is used to combine each data type's score into a single joint likelihood. First, the scores are standardized to have mean 0 and standard deviation 1, with $log(\tilde{z}_{ik}) = log(z_{ik}) - \mu_k / \sigma_k$ (where z is a stand-in for the expression (r), sequence (s), or association network q^n p-value). This is done so that one type of score doesn't overpower the others simply because it tends to have better p-values than the others. Then, we combine the scores into a single value, on which we will perform the regression:

$$g_{ik} = r_0 \log(\tilde{r}_{ik}) + s_0 \log(\tilde{s}_{ik}) + \sum_{n \in N} q_0^n \log(\tilde{q}_{ik}^n), \tag{2.4}$$

where r_0, s_0, and q_0^n are the weights that are specified by the current annealing iteration for the expression, sequence, and network scores, respectively. The idea here is to weight each type of p-value based on how important we think it is given where in the annealing process we are.

The constrained logistic regression is then defined as:

$$\pi_{lk} \equiv p(y_{lk} = 1 | X_k, S_i, M_k, N) \propto \exp(\beta_0 + \beta_1 g_{lk}), \tag{2.5}$$

where l is used to define a gene *or* experiment (replacing i and j from before). π_{ik} is then the likelihood that a given gene or experiment l belongs to bicluster k.

cMonkey puts all of these methods together into a single iterative procedure. (i) Create random biclusters using a random initialization method for each bicluster, and start at a high annealing temperature. (ii) Calculate the joint likelihood that each gene/experiment belongs in each bicluster. (iii) For each gene / experiment add or drop it from each bicluster according to the probabilities:

$$p(\text{add}|\pi_{lk}) = e^{-\pi_{lk}/T}; p(\text{drop}|\pi_{lk}) = e^{-(1-\pi_{lk})/T}, \tag{2.6}$$

where T is the current annealing temperature or until the maximum number of moves per iteration is achieved (the default value is 5) and π_{lk} is the likelihood that a given gene or experiment l belongs to bicluster k, as defined in Equation (2.5). (iv) Lower the annealing temperature, then repeat steps (ii) - (iv) until the temperature reaches 0 or its minimum value. By default, the annealing temperature begins at 0.15 and goes down in even steps to 0.05 over 100-150 iterations.

The result of this procedure yields biclusters of genes and experiments. These data can be used to create "super genes" for use in a network inference algorithm. By taking the list of genes in a bicluster, we can then average those values together to create the expression values over experiments for this gene. The idea behind this is that if these genes are part of the same pathway and behave like each other, then we can reduce the amount of noise and variance in the expression measurements by averaging their values together.

2.2.4 Walkthrough Example on Toy Data

We will now demonstrate cMonkey in action using a small toy example. Our example will contain only 5 genes and 5 experiments. We begin by randomly selecting one of the five different methods for seeding a bicluster. Let's assume that we choose method 1: selecting a single random gene to start a bicluster and begin with an annealing temperature of 0.15. Gene 3 is selected to begin the bicluster. We then calculate the joint likelihood that each gene or condition belongs in this bicluster. Table 2.1 represents the joint probability (the result of Equation (2.4)) that each gene/experiment belongs in a bicluster containing only gene 3.

	Experiment 1	Experiment 2	Experiment 3	Experiment 4	Experiment 5
Gene 1	0.8	0.2	0.01	0.3	0.02
Gene 2	0.3	0.9	0.1	0.1	0.6
Gene 4	0.5	0.6	0.4	0.3	0.2
Gene 5	0.2	0.1	0.7	0.2	0.3

Table 2.1 Score of each gene/experiment with respect to gene 3.

We can see from Table 2.1 that there are some gene/experiment combinations that have a good chance of actually belonging to this bicluster. Specifically, Gene 1/Experiment 1 and Gene 2/Experiment 2. Given the joint probabilities in Table 2.1, we calculate the probability of each gene/experiment being added to the bicluster using the left side of Equation (2.6). This takes into account the current annealing temperature, which is set to 0.15 at the beginning. This gives us Table 2.2, which shows $1 - $ score.

	Experiment 1	Experiment 2	Experiment 3	Experiment 4	Experiment 5
Gene 1	0.99	0.73	0.07	0.86	0.12
Gene 2	0.86	0.99	0.49	0.49	0.98
Gene 4	0.97	0.98	0.93	0.865	0.74
Gene 5	0.74	0.49	0.99	0.73	0.86

Table 2.2 $1 - $ Score of each gene/experiment with respect to the new gene 3 bicluster. The numbers are high because the annealing temperature is high. As the temperature cools, these scores will decrease.

Gene/experiment values with high joint probabilities in Table 2.1 have correspondingly high numbers in Table 2.2. To decide whether or not to add a particular gene/experiment gx to the bicluster, we select a random number between 0 and 1. If that random number is less than the value shown in Table 1.2 for gx, then we add gx. Otherwise, don't add gx to the bicluster. For example, suppose we select a random number of 0.6 for the value of gene 1/experiment 1. Because the random number is less than the listed value, we add gene 1/experiment 1 to the bicluster. By contrast, if, for gene 1/experiment 2, we select a random number of 0.9. We do not add that gene/experiment combination to the bicluster.

A similar process is then carried out to decide whether or not to drop gene/experiment pairs from the bicluster, using the right side of Equation 2.6 instead of the left. Once this process has been carried out for each bicluster (in this example, there is only one), the annealing temperature is dropped. Dropping the annealing temperature af-

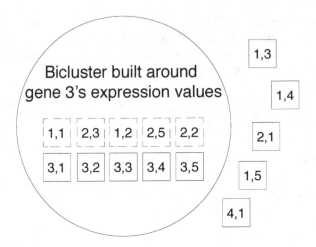

Fig. 2.1 A figure showing an example bicluster that was seeded with the expression values of gene 3. The observations are labeled as boxes, with the numbers representing "gene number, experiment number". Dashed boxes are the 5 freshly added observations, and solid boxes indicate observations that were already part of the bicluster. Shown outside of the bicluster are some other observations that did not make the cut.

fects the results of Equation 2.6, making it less likely that a move is made. This process continues until either the minimum annealing temperature is reached or until a maximum number of moves (additions or deletions) is made for this turn, yielding a collection of biclusters.

Parameter Name	What it does	Default Value
m	Maximum number of moves per iteration	5
T_{max}	The starting annealing temperature	0.15
T_{min}	The ending annealing temperature	0.05
T_{step}	The number of steps between T_{max} and T_{min}	100

2.3 Factor Analysis for Bicluster Acquisition (FABIA)

2.3.1 What it Does

Factor Analysis for Bicluster Acquisition (FABIA) [14] biclusters genes and experiments using Factor Analysis. Factor Analysis takes a set of data (in our case, the expression values of genes in experiments) and explains them in terms of an often smaller set of parameters called factors. In a non-genetic context, consider lung cancer data about people. Imagine that the data are rows of people with lung cancer and columns are indirect data that are correlated with a direct cause of the cancer (age, socio-economic status, location, etc.). What FABIA tries to do is explain the underlying relationship between indirect data. For example, a direct cause of lung cancer such as asbestos exposure (i.e., a "factor") may be partially explained by a combination of other features correlated with the direct cause, such as whether or not a person lived in a time and area where asbestos exposure was common. FABIA uses Expectation-Maximization [6] to generate the biclusters. Biclusters are then ranked by mutual-information content, and weaker members of each bicluster are optionally pruned with a threshold.

2.3.2 The Data

FABIA uses steady-state data. The experiments can be genetic or external perturbations, and should be normalized to have mean 0 and standard deviation 1. Time-series experiments may also be used, but they will be treated as individual steady-state experiments.

2.3.3 The Strategy

FABIA tries to find a set of factors z that explain observed expression values in X. To do this, we need to find a good set of weights called "factor loadings" that connect a factor in z to the observation in X. Which genes are part of a given factor z_i is decided by the factor loadings in λ associated with z_i. We also want to model the measurement noise ε of each observation and then remove that noise in order to calculate an "idealized" expression value (see Figure 2.2).

Formally, this can be modeled as:

$$X = \sum_{i=1}^{p} \lambda_i \tilde{z}_i + \varepsilon = \Lambda \tilde{z} + \varepsilon, \tag{2.7}$$

where ε is additive noise, p is the number of biclusters, λ_i is a sparse vector of factor loadings, and $\tilde{z}_i$ is the ith value in a vector of $\tilde{z}$ factors. The approach to fitting this model uses some advanced techniques.

For this model, we want to find the parameters Λ and Ψ that best explain the data. $\Psi = \text{Cov}(\varepsilon)$ is a matrix that represents the covariance of the noise of the expression values in X. Λ (the factor loadings) represents the connections between factors in z and observations in X. We find parameters to best explain X using the Expectation-Maximization (EM) [6] algorithm.

Expectation-maximization (EM) is an iterative method for finding the maximum likelihood of a set of parameters. FABIA uses a special kind of EM algorithm group

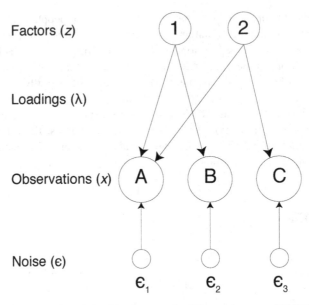

Fig. 2.2 A figure showing the relationships between the different elements of FABIA. In this case, the two factors (z) explain the three observations (x) through factor loadings (λ). Factor 1 connects observations A and B. Factor 2 connects observations A and C. FABIA treats these factors as biclusters, giving us 2 biclusters where bicluster 1 contains A and B, and bicluster 2 contains A and C.

called "variational EM" [10] [24]. This implementation of the variational EM algorithm places a Laplacian prior on the problem to enforce sparsity. This is similar to the constraint that l_1 optimization uses: non-zero values are implicitly penalized, so "weak" connections will quickly drop to zero, yielding a more parsimonious result. For variational EM, this is done by optimizing for the variational parameters ξ that are part of the Lagrangian dual formulation of the problem. By optimizing the variational parameters, we obtain a lower bound on the likelihood of the model. The goal is to find a maximum value for this lower bound. FABIA uses variational EM to search for the combination of Λ and Ψ that fits the data the best. This works in two steps. First, the expectation (E) step calculates the expected log-likelihood of the current parameters, i.e., how likely it is that the current parameters fit the data better than a null model. Next, the maximization (M) step computes a new set of model parameters Λ and Ψ that maximize the expected log-likelihood from the previous E-step. These two steps are repeated until a maximum number of iterations is reached or the model is changing too slowly.

Once we've obtained a good estimate of the parameters Λ and Ψ, we can rank the resulting biclusters according to how much information each contains about the data. We do this by calculating the mutual information between the data X and the factors of each bicluster z_i^T. The information content of a bicluster will grow with the number of nonzero values (i.e., the size of the bicluster) in each λ_i, so in general, the larger the biclusters are then the more information about X they contain. Finally, each bicluster may optionally be pruned by taking the absolute value of the factors and factor loadings for each bicluster, and selecting only the values that are above a certain threshold.

2.3.4 Walkthrough Example on Toy Data

FABIA works by iteratively adjusting the loading matrix Λ and the covariance matrix of the noise Ψ. Imagine that we are working with a very small dataset of only 9 observations (3 genes, 3 experiments). We are looking to fit these 9 observations into 3 factors (biclusters). Λ is then a 3x9 matrix initialized randomly to values between 0 and 1. Ψ is initialized to be:

$$\Psi = \text{diag}(\text{CoV}(x) - \Lambda\Lambda^t). \tag{2.8}$$

The variational parameters ξ are initialized to be 1.

Our initial guess, a randomly generated matrix, is a mess. It is likely that it is fully connected, i.e., there is an edge from each observation to each bicluster. This is obviously not a useful result, so we want to begin pruning the edges that don't belong. We then enter the E-step of the EM algorithm, where we calculate the log-likelihood that the current parameters Λ and Ψ fit the data better than a null model. The likelihood that each observation x_i is a factor z_j is calculated. Obviously, at this point that, likelihood is going to be very low, as we have a random model right now.

Next, we use this likelihood in the M-step in order to find new matrices Λ and Ψ that better explain the model. To update Λ, a convex quadratic problem is solved. The basic idea though is that we use the likelihood from the E-step that each observation in X belongs to a given factor in z, given the rest of the observations currently in that factor. The factor loading in Λ^{new} is then updated to reflect the likelihood that an observation x_i is in factor z_j. Small values of factor loadings are penalized and forced to 0. Then, Ψ is updated using the updated Λ^{new}. The basic idea behind updating Ψ is to use the new information available from Λ^{new} and from the likelihoods calculated during the E-step to estimate the variance in the observations X. If an observation strongly belongs into a bicluster, then maybe that gene really does have a very high or very low value, and it isn't simply a noisy observation. In this case, the value in Ψ is reduced.

At the end of the M-step we have a slightly better guess as to which genes actually belong to each bicluster. For our example, let's say that Λ now looks like Table 2.3. We can see that observation 8 seems to strongly belong to factor 1, and that observation 5 strongly belongs to factor 2. There are also several small values. These are likely to drop to 0 in following iterations as their small values are penalized. As more EM iterations are performed, these numbers will slowly change until either the numbers converge to values and stop changing or we hit the maximum number

	Obs. 1	Obs. 2	Obs. 3	Obs. 4	Obs. 5	Obs. 6	Obs. 7	Obs. 8	Obs. 9
Factor 1	0	0.5	0.3	0.1	0	0.2	0.9	0.4	0.2
Factor 2	0.1	0	0	0.9	0.5	0	0.2	0.1	0.6
Factor 3	0.2	0.1	0	0.4	0.3	0.1	0	0.5	0.3

Table 2.3 Example Λ matrix after 1 Expectation-Maximization step.

of iterations. The matrix Λ is then a map between our factors and observations, and this map is used to create the biclusters.

Parameter Name	What it does	Default Value
p	Maximum number of biclusters	Depends on data
cyc	Maximum number of iterations	500

Chapter 3
Step 2: Use Steady State Data for Network Inference

3.1 Introduction

We say an organism is in steady state if the values of what we are measuring (e.g. gene expression) won't change unless we change its conditions in some way. For example, the organism may be in one steady state in a low nutrient condition, another in a high nutrient condition, and still another after some mutation has occurred and we have waited until transient effects have died out. Steady-state data can arise from experiments in which one or more genes have been knocked out, overexpressed, or otherwise perturbed. If you know what happens to the network when a gene is missing or when it has been perturbed, it is easier to infer which genes it influences.

The following algorithms use steady-state expression data to infer networks. Which algorithm to choose depends partly on the data that is available. We suggest the "sweet spot" for each technique below.

3.2 Median-Corrected Z-Scores (MCZ)

3.2.1 What it Does

Median-Corrected Z-Scores (MCZ) [11] tied for first place in the DREAM4 network topology inference challenge when mutation data but no time series data was available. When both were available, then MCZ could be used as a one step in a workflow as we explain in chapter 5. The basic idea is that if gene j influences i, then knocking out j should change the value of i in a significant way. Whether or not the knockout of gene j affects gene i is based on the number of standard deviations the expression value of i is from its median value across all experiments. The further away it is, the more likely that gene j has an effect on it.

J. M. Lingeman and D. Shasha, *Network Inference in Molecular Biology,*
SpringerBriefs in Electrical and Computer Engineering,
DOI: 10.1007/978-1-4614-3113-8_3, © The Author(s) 2012

Rating: DREAM Best

3.2.2 The Data

For this challenge, two types of data were available for a 100 gene network: (i) wild-type observations for each gene and (ii) a knockout dataset where there is one experiment corresponding to each gene in which that gene is knocked out. The DREAM4 in-silico challenge had a relatively small wild-type dataset with only 11 expression values for each gene. As explained in section 1, the DREAM datasets are simulated datasets having similar noise statistics to real biological datasets but with known correct answers that are hidden from the algorithm under test. The correct answers are in the form of coefficients of positive or negative influence of one gene on another.

3.2.3 The Strategy

In order to get a robust value for the median expression value of each gene, the algo-rithm combines the wild type measurements of each gene with the gene's expression value in the knockout data set. The idea here is that the knockout of a gene g' should affect the expression values of only a few other genes, leaving the vast majority un-perturbed. Hence, in the majority of the knockout experiments, a particular gene g is unaffected by the knockout. The median of the combined dataset is used as an estimate of the wild-type (non-knockout) population median. The median is used because it is more robust to outliers than the mean. This value is represented by x^{wt}.

In the DREAM4 data and in actual expression data, the noisiness of the data is a function of the gene's expression. Often, the higher the expression value, the noisier the data. The z-score normalizes for this effect at least to some extent. If we knock out a gene x_j that had an edge to gene x_i, we would expect to have the value of x_i change substantially in one direction or the other, moving us away from the median. To calculate whether a gene x_i is a target of a transcription factor (TF) x_j, we calculate the z-score with respect to the median of x_i for the x_j knockout experiment

$$z(x_i | x_j^{ko}) = \frac{x_{ij}^{ko} - x_i^{wt}}{\sigma_i}, \tag{3.1}$$

where x_{ij}^{ko} is the value of x_i given that x_j is knocked out, x_i^{wt} is the median of the wild-type and knockout datasets for x_i (the population median), and σ_i is the standard deviation of the wild-type and knockout datasets for x_i. By calculating the MCZ for each combination of gene i with gene j removed, we can obtain a ranking of regulatory interactions that can be used to build a network topology. Thus, j ranks higher than k in its influence on i if knocking out j yields a greater Z score for i than the Z score for i when knocking out k.

3.2.4 Performance on Examples

The performance of MCZ was tested on the example small 10 gene network and large 100 gene network from the DREAM4 In-Silico Challenge. The DREAM4 challenge provided a very complete dataset, including knock-out, knock-down, time-series, and wild-type data. MCZ performs extremely well on the small 10 gene network. The area under the receiver-operator characteristic curve (AUROC) is 0.90.

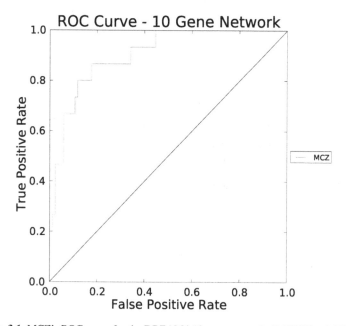

Fig. 3.1 MCZ's ROC curve for the DREAM4 10 gene network. (AUROC = 0.90)

The ROC curve in Figure 3.1 is a good indicator that the edges with the highest scores are in fact likely to be real edges. The precision-recall curve also shows good performance with an area of 0.65. What this precision-recall curve tells us is that the highest ranked edges from MCZ are very accurate, and then we start falling into a "saw tooth" pattern, where a false guess is made, followed by a correct guess. In fact, this graph tells us that roughly 20% of the true positives appear without a single false positive (a guessed edge that is not a real edge). About 60% of MCZ's guesses are correct by the time it has recovered 60% of the true edges. This is extremely good performance.

Not all real world datasets are as complete as this simulated DREAM4 dataset, so what happens if we use less data? The results from table 3.1 are roughly the same as long as we have knock-down and/or time-series data contributing to the wild-type median. However, if we use only the provided wild-type data to estimate

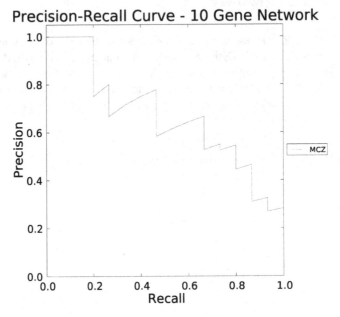

Fig. 3.2 MCZ's Precision-Recall curve for the DREAM4 10 gene network. (AUPR = 0.65)

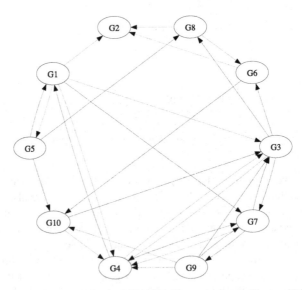

Fig. 3.3 MCZ's inferred network for the DREAM4 10 gene network. The top 25% of the ranked edges were used. Solid lines are correct guesses, dotted lines are missed edges, and dashed lines are incorrect guesses. In the top 25% of edges, MCZ misses some of the edges involving G3.

	Full Data	No KD	No TS	No KD & No TS
AUROC	0.90	0.89	0.90	0.77
AUPR	0.65	0.63	0.64	0.50

Table 3.1 Table of areas under the receiver-operator curve (AUROC) and the precision-recall curve (AUPR) for MCZ for the entire small 10 gene dataset, when knock-down (KD) data are removed, when time-series (TS) data are removed, and both knock-down and time-series data are removed. All runs include the knock-out and wildtype datasets. The results are roughly the same as long as either knock-down or time-series data are used. However, if neither is used, MCZ does not perform as well.

the median, MCZ does not perform as well. The reason is that the wild-type data provided are not an accurate estimate of the actual median expression value, and we cannot obtain a good estimate without at least one of the other datasets.

MCZ works well on a small dataset, but how does it work on a larger dataset? MCZ scales well to the 100 gene network from the DREAM4 challenge (Table 3.2).

	Full Data	No KD	No TS	No KD & No TS
AUROC	0.91	0.90	0.91	0.90
AUPR	0.45	0.49	0.45	0.48

Table 3.2 Table of areas under the receiver-operator curve (AUROC) and the precision-recall curve (AUPR) for MCZ for the entire large 100 gene dataset, when knock-down (KD) data are removed, when time-series (TS) data are removed, and both knock-down and time-series data are removed. All datasets include knock-out and wild-type data. The extra data help less for the larger dataset than for the smaller one.

For this dataset there was no substantial difference between using the full dataset to estimate the median expression value of each gene or only part of it. This suggests that the provided wild-type dataset provides a good estimation of each gene's median expression values, and that little was gained by adding more information.

On both of the test datasets, MCZ performs very well. It also scales well. The running time of the algorithm on both the small and the large network is under 30 seconds. If you have both a full set of knock-out data and wild-type expression data, MCZ should be used.

3.3 Network Identification by Multiple Regression (NIR)

3.3.1 What it Does

Network Identification by Multiple Regression (NIR) [9] uses multiple regression
to infer networks from steady state expression data resulting from a known initial
perturbation. The basic assumption is that a network of genes can be approximated
by a system of linear equations:

$$dX/dt = AX + U \qquad (3.2)$$

where X is an n by m matrix of steady-state expression data. In X, each column
represents an experiment and the rows represent genes. A is an n by n normally dis-
tributed matrix representing the network model. This matrix representation implies
that each gene's expression is a linear sum of a row of coefficients from A and gene
values from X. U is an n by m matrix representing the degree to which the gene that
is the target of a perturbation is perturbed in each experiment (as a value between 0
and 1). For example, if the gene is fully knocked out, it may have a value of 1. A
gene that is knocked-down may have a value of 0.5. Genes that have not been experimentally perturbed have values of 0. dX/dt represents how much the expression
data are changing per unit of time. Since NIR is used with steady-state data (that is,
the data are assumed to change little over time), dX/dt is set to 0. Thus, the above
equation reduces to:

$$-U = AX \qquad (3.3)$$

Multiple regression is used to select a promising network model A.

Rating: DREAM Fair

3.3.2 The Data

NIR infers networks from steady-state data. Each gene is assumed to have been
perturbed in at least one experiment. These perturbations must be relatively "small".
Gardner, et al. [9] define "small" as being an experiment that does not knock the
network out of its steady-state basin of attraction. Intuitively, this means that the
perturbation should be small enough such that the linear approximation works. In
practice, this means try NIR if you have perturbation data and hope for the best.

3.3.3 The Strategy

The U matrix is an n by m matrix that marks which genes were perturbed in each experiment. In the simplest case, U is a binary matrix, where a 1 at position i, j indicates that gene i was perturbed in experiment j. The A matrix is an n by n matrix that holds the network estimation. This is the matrix we are trying to find. X is the n by m matrix of expression values, where there are n genes and m experiments. We can think about the above equation more concretely by taking just one row a_i from A and one column x_j from X, solving for u_{ij} of U. We want to find a combination of values in a_i that, when multiplied by x_j and added together, equals $-u_{ij}$. Since there are an extremely large number of possible solutions, we need some way to select the best answer.

NIR picks the solution by creating a multiple linear regression model. A multiple regression is a model that can account for more than one independent (predictor) variable. For example, consider estimating the price of a house. Example independent variables that a house price model might include are: number of bedrooms, number of bathrooms, age of the house, and size. We can then build a multiple regression model from housing data. Building this model gives an estimate of how much each of these predictor variables influence the price. It also allows us to estimate the price of a new home, given the predictor variables. NIR applies this to each experiment. The independent variables are each possible set of k out of n genes. k is a user-defined parameter that enforces sparsity in A limiting the maximum number of dependencies each "target" gene can have on other genes. The dependent variable is the negative perturbation ($-u_{ij}$) value for the target gene/experiment. These steps are repeated for each gene/experiment combination. The network matrix A is built from the model weights that best predict each gene.

Specifically, NIR uses least squares regression. Least squares regression attempts to minimize the sum of squares cost function:

$$SSE_i^k = \sum_{l=1}^{M} (y_{il} - \hat{b}_i^T \cdot z_l)^2 \qquad (3.4)$$

where k represents the set of genes being examined, i is the target gene, l is the current experiment, y_{il} is the negative perturbation value for gene i in experiment l, $\hat{b}$ are the model weights for gene i, and z_l^k are the expression values for the currently selected set of k genes in experiment l. The basic idea is to choose the weights $\hat{b}$ that minimize the sum of squared errors. Intuitively, the squared error measures the difference between how much the target gene was perturbed and the perturbation that the current model represents. For example, if the gene and the experiment that are currently being analyzed have a perturbation value in U of 1, then ideally we'd like to find a set of k weights $\hat{b}$ (where at least one weight is non-zero) whose dot product with the current experiment's expression values is equal to -1 (since $y_{il} = -u_{il}$), making the error 0. The source nodes of the edges having non-zero weights correspond to the genes that regulate the target gene i.

For each gene, the model with the smallest sum of squared error is then tested for significance using a F-Test:

$$F = \frac{(SSE_0 - SSE_k)/k}{SSE_k/(m-k)} \tag{3.5}$$

where SSE_0 is the sum of squared errors when the weights $\hat{b}$ are set to 0. The F-test is comparing the error in the inferred model to the error from the null hypothesis (where the model is set to 0). If the F-Score is higher than a threshold F*, our model fits the data significantly better than the null hypothesis. The F* threshold is the value of an F-distribution with k and $m - k$ degrees of freedom at a desired confidence level. Finally, we can use the model weights to fill in the target gene's connections in the network matrix A, giving us our network.

A practical limitation of NIR is its runtime. $\binom{n}{k}nm$ multiple regressions must be run to exhaustively cover all possible solutions. This severely limits the usefulness of NIR for even relatively small networks. Recently, Gregoretti, et al. [12] implemented a parallelized and optimized form of NIR to alleviate this problem. The major speedup comes from running the multiple regressions in parallel with each other across multiple processors. Since each regression can be run independently of all of the others, the overall runtime of NIR can be cut in proportion to the number of processors available to run it. The result is still exponential of course, but the speedup increases the sizes of the networks that can be modeled.

Parameter Name	What it does	Default Value
k	Maximum number of edges any gene can have to others	5

3.3.4 Performance on Examples

NIR was tested using the 10-gene DREAM4 network and a DREAM4 100-gene network. NIR's main parameter is called k, which controls the maximum number of connections each gene is allowed to have.

NIR only achieved rather poor performance on the small network. It has a somewhat low area under the ROC curve (0.66 at $k = 5$), and a low area under the precision-recall curve (0.23 when $k = 3$).

These results are interesting. The parameter k does not seem to have much effect across different runs of NIR. The reason is that the component of NIR that is enforcing sparsity quickly switches to Least Angle Regression rather than the parameter k. While it is good news that NIR is not extremely sensitive to its parameter k (at least in the large network), it does not perform very well. The area under the ROC curve is only about 0.62 for each value of k, and the area under the precision-recall curve is about 0.07.

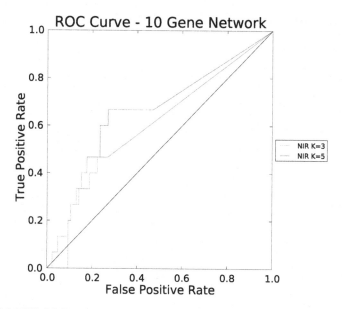

Fig. 3.4 NIR's ROC curve for the DREAM4 10 gene network at different values of k, the parameter that describes how many connections each gene can have. (Best AUROC = 0.66 at $k = 5$.)

There is a possible reason for this poor performance: not enough combinations of edges were tried. This implementation of NIR uses a heuristic to try to select the best combinations of edges iteratively, instead of exhaustively trying each possible combination. It is simply not feasible to try every possible combination of genes on a 100 gene network without severely limiting the number of edges each gene can have. Even then, the number of regressions to perform grows too quickly to remain feasible on larger networks.

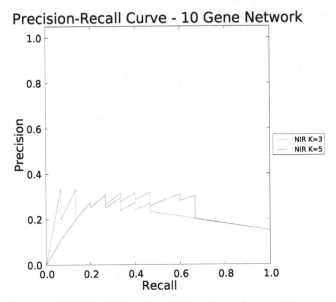

Fig. 3.5 NIR's precision-recall curve for the DREAM4 10 gene network at different values of k, the parameter that describes how many connections each gene can have. (Best AUPR = 0.23 at $k = 3$.)

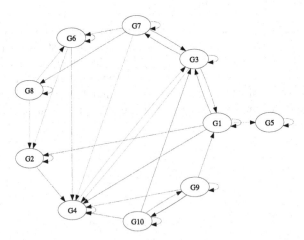

Fig. 3.6 NIR's inferred network for the DREAM4 10 gene network. The top 25% of the ranked edges were used. Solid lines are correct guesses, dotted lines are missed edges, and dashed lines are incorrect guesses. NIR infers only six correct edges in the top 25%, and misses all of the edges around G1 and G10.

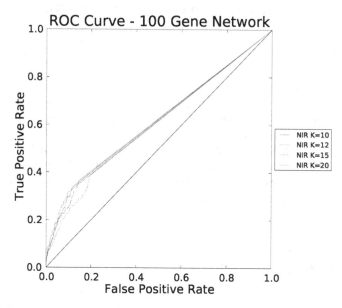

Fig. 3.7 NIR's ROC curve for the DREAM4 100 gene network at different values of k, the parameter that describes how many connections each gene can have. (AUROC is about 0.62 for most values of k.)

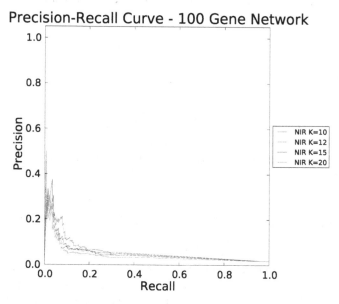

Fig. 3.8 NIR's precision-recall curve for the DREAM4 100 gene network at different values of k, the parameter that describes how many connections each gene can have. (AUPR is about 0.07 for most values of k.)

3.4 Gene Network Inference with Ensemble of trees (GENIE3)

3.4.1 What it Does

GENIE3 is an algorithm that uses an ensemble of regression trees to infer networks from multifactorial data. Multifactorial data are created by perturbing some or all of the genes in a dataset (e.g. by providing a nutrient) and then taking a measurement after the organism has reached steady state. This is in contrast to the steady state data we've used previously, where only one perturbation was applied at a time. GENIE3 was the best performer in both the DREAM4 and DREAM5 *In Silico* Multifactorial Challenges. These challenges used only multifactorial data to create a network, as opposed to the single gene knockout experiments used in many of the other challenges.

Rating: DREAM Good

3.4.2 The Data

GENIE3 is the preferred method for multifactorial data, though it can also be used with any other kind of steady-state data or time-series data (by treating the time-series data as steady-state).

3.4.3 The Strategy

The GENIE3 algorithm operates on expression data where each gene's expression has been normalized to unit variance (i.e., variance of 1) and mean 0. The algorithm works in three steps: First it creates an ensemble of regression trees for each gene in the network. Next, possible regulators are ranked from each regression tree. Finally, it ranks the inferred edges overall.

3.4.3.1 Step 1: Regression Trees

GENIE3 creates a regression tree to predict the behavior of each gene g in the dataset. Regression trees work by recursively splitting the dataset into ever smaller subsets. The nodes of the regression tree will split the dataset based on values of genes other than g. To avoid ambiguity with the term node in the final result, we will call the regression tree nodes decision points. Each decision point splits the dataset into two sub-datasets such that each sub-dataset has a small variance in the target gene's expression values. The split is based on a single gene x other than g

and a threshold value for x. The idea is that if gene x causes a split in the regression tree for target g, then there is a potential causal gene from x to g. We then look at x's expression value in each experiment (where the experiment may be a perturbation experiment on x or on any other gene or simply a treatment experiment on the whole organism). If the expression value of x in the experiment is above the threshold, the experiment goes into one group. If the expression value is below the threshold, the expression goes into the other group. This process is continued recursively on the sub-datasets until no more splits can be made (i.e., it is fully split).

As an example, consider the following table:

Experiments	Genes			
	G1	G2	G3	Target
E1	0.4	0.8	0.4	0.5
E2	0.3	0.2	0.3	0.9
E3	0.5	0.3	0.7	0.8

Table 3.3 Example data for GENIE3. The idea behind the GENIE3 algorithm is to split this data into two groups of experiments that have minimal variance on the target gene.

We want to split the experiments into two groups that minimize the variance of the target gene's expression values. We can see that the ideal split is to have experiment E1 alone in one group, and E2 and E3 in the other group. For this purpose, G2 is the best candidate. Since G2's value in E1 can be cleanly split from its value in E2 and E3, we can select 0.5 as a threshold (as it is between G2's E1 and E3 values). Any values above 0.5 go to group 1, and any values less than or equal to 0.5 go to group 2.

Group	Experiments	Genes			
		G1	G2	G3	Target
Group 1	E1	0.4	0.8	0.4	0.5
Group 2	E2	0.3	0.2	0.3	0.9
	E3	0.5	0.3	0.7	0.8

Table 3.4 Example data for GENIE3 from 3.4.3.1 with the split inserted as the thick line between experiments. Splitting the dataset into groups 1 and 2 minimizes the variance of each of the groups.

When G2's expression value is above 0.5, the target gene has low expression values. When G2's value is below 0.5, the target gene has high expression values. Thus, we have identified a potentially casual edge (that G2 has a repressive effect on the target gene).

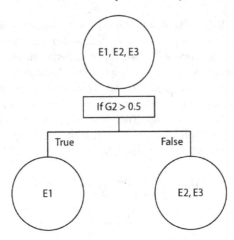

Fig. 3.9 The tree after the first split. The top circle represents the entire dataset. The rectangle represents the decision node, containing the criteria of the split. The child circles contain their respective experiments after the split.

3.4.3.2 Step 2: Selecting the split using Random Forests

In order to find splits that are robust to slight changes in the data, Random Forests [4] are used. Random Forests use bootstrapping and random feature selection to reduce variance across the dataset by averaging predictions. For each tree in a Random Forest, a bootstrap training set of about 2/3 the size of the original dataset is created by random sampling of the set of experiments with replacement. The tree is then built by taking K random splits for each decision node. From the Random Forest literature, K is usually defined as $K = \sqrt{p-1}$ or $K = p-1$ where p is the number of potential regulators (e.g. transcription factors) if known. The decision split is the randomly chosen split that most reduces the variance of the target gene's expression values.

An importance score is then calculated for each decision point in the tree:

$$I(N) = \#SVar(S) - \#S_tVar(S_t) - \#S_fVar(S_f) \qquad (3.6)$$

where N is the current decision point being evaluated, S is the subset of experiments that are below decision point N in the tree, S_t and S_f are the subsets of experiments on the true and false branches of decision point N, respectively, $Var(.)$ is the variance of the target gene in a subset, and $\#$ denotes the number of experiments in its associated subset. This importance score is a measure of how much variance is explained by splitting the dataset on the decision point's gene and threshold. Intuitively, it can be read as "how much is the variance of the dataset at decision point N reduced by subtracting out the variance of each of the subsets, weighted by the number of experiments in each set?" If the score is high, than that means that the variance is substantially reduced and this gene might regulate the target gene (as in the example shown above from step 1). If the score is low, then the split did not reduce much variance, and this gene probably does not regulate the target gene.

3.4.3.3 Step 3: Ranking possible regulators from each tree

Once a tree for a gene g is created, we can rank the influence of every other gene on g. A score for a potential regulator gene g' is calculated by summing all of the importance scores from the nodes where g' was selected for splitting. Genes that are never selected for splitting are given scores of 0. We can then rank the scores to determine which genes g' were most important for regulating gene g.

3.4.3.4 Step 4: Ranking the inferred edges

In [15], 1000 Random Forests are created. Importance scores are generated for each tree in each forest, giving a list of potential regulators for each tree. The trees belonging to each target gene are then grouped together, and the importance scores for potential regulators are then averaged together. These averaged scores can then be used to rank potential regulators for each gene.

Parameter Name	What it does	Default Value
K	Number of splits to test per node	[15] uses $\sqrt{p-1}$ where p is the number of transcription factors (if known), otherwise $N-1$.

3.4.4 Performance on Examples

GENIE3 was tested using the 10-gene DREAM4 network and a DREAM4 100-gene network. Here, we tested GENIE3's results when using different amounts of data. GENIE3 won the DREAM4 multifactorial challenge. This means that GENIE3 performed better than any other algorithm using only multifactorial (MF) data. We tested GENIE3 by adding the different DREAM4 datasets one at a time.

GENIE3 performed very well on the small 10-gene network. Using only the multifactorial data, GENIE3 gave impressive results (AUROC = 0.71, AUPR = 0.30). As expected, when the other data are added, GENIE3 performs much better (Max AUROC = 0.86, Max AUPR = 0.43).

Performance on the ROC curve remains fairly strong in the 100 gene network using all of the datasets (AUROC = 0.70), but the precision-recall curve is weak (AUPR = 0.05). This means that, overall, the precision (proportion of correct guesses to total number of guesses made so far) is low compared to the recall

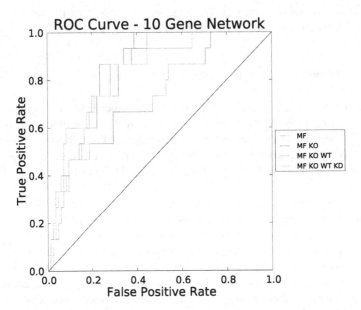

Fig. 3.10 GENIE3's ROC curve for the DREAM4 10 gene network using multifactorial (MF), knockout (KO), wildtype (WT), and knockdown (KD) datasets. Best performance (AUROC = 0.86) was achieved with only multifactorial and knockout datasets.

(proportion of correct guesses to the total number of real edges), while the recall compared to the false positive rate (proportion of false positives to total number of edges) remains high. In short, GENIE3 is pretty good at guessing edges, but it can make quite a few incorrect guesses for each correctly inferred edge. GENIE3 should be used when inferring networks from multifactorial data.

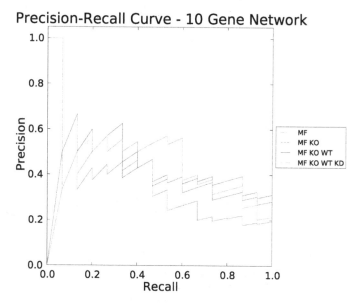

Fig. 3.11 GENIE3's precision-recall curve for the DREAM4 10 gene network using multifactorial (MF), knockout (KO), wildtype (WT), and knockdown (KD) datasets. Best performance (AUPR = 0.43) was achieved when using all datasets.

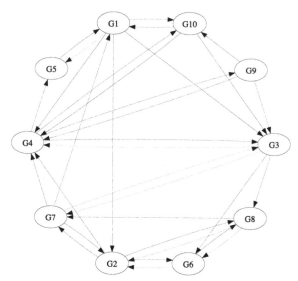

Fig. 3.12 GENIE3's inferred network for the DREAM4 10 gene network. The top 25% of the ranked edges were used. Solid lines are correct guesses, dotted lines are missed edges, and dashed lines are incorrect guesses. In the top 25% of edges, GENIE3 is unable to correctly infer G1 and G10's relationships to other genes.

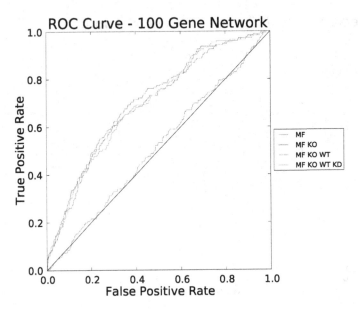

Fig. 3.13 GENIE3's ROC curve for the DREAM4 100 gene network using multifactorial (MF), knockout (KO), wildtype (WT), and knockdown (KD) datasets. Best performance (AUROC = 0.70) was achieved using all of the datasets.

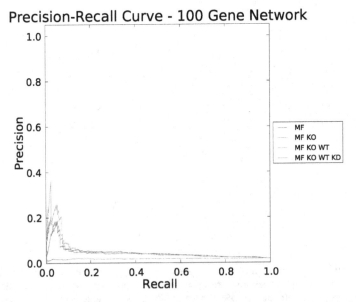

Fig. 3.14 GENIE3's precision-recall curve for the DREAM4 100 gene network using multifactorial (MF), knockout (KO), wildtype (WT), and knockdown (KD) datasets. Best performance (AUPR = 0.05) was achieved when using all datasets.

3.5 Context Likelihood of Relatedness (CLR)

3.5.1 What it Does

The Context Likelihood of Relatedness (CLR) [8] algorithm uses mutual information to infer networks from steady-state data. Mutual information is a measure of
dependence between two random variables. In this case, mutual information is used
as a metric of similarity between two gene expression profiles. The intuition is that
if genes g and g' are connected by an edge, then the expression value of g gives us
some information about g' (i.e. g's expression value affects the expression value of
g'). If g and g' have no edge, then knowing the expression value of g tells us nothing about the expression of g'. The mutual information score indicates how much
information is shared between the genes. A score of 0 indicates that there is no
information shared between two genes.

The basic idea behind CLR is to form a matrix of mutual information scores by
calculating the mutual information between each pair of genes in the network. These
scores are then compared to a background distribution and a z-score is calculated. A
high z-score indicates a potential mutual information edge in the network between
the two genes, while a low z-score means that there probably isn't an edge. CLR can
output only undirected edges due to the bidirectional nature of mutual information.
If transcription factors are known, then we can guess the directionality when exactly
one of the genes is a transcription factor.

Rating: DREAM Good

3.5.2 The Data

CLR uses either over-expression or knockout steady-state experiments to infer a
network. Recall that over-expression data can be obtained by mutating an organism so that certain genes have greater expressions than they would naturally. An
over-expression mutant for g can either produce excess amounts of g throughout its
lifetime or only when the experimentalist wishes. For the purpose of this descrip-
tion, we don't distinguish the two possibilities.

3.5.3 The Strategy

First, the mutual information between each pair of genes is calculated:

$$I(X;Y) = \sum_i \sum_j p(x_i, y_j) \log \frac{p(x_i, y_j)}{p(x_i)p(y_j)} \qquad (3.7)$$

where $p(x_i, y_j)$ is the joint probability density function between X_i (the expression profile of gene$_i$) and Y_j (the expression profile of gene$_j$). $p(x_i)$ and $p(y_j)$ are the marginal probability density functions of X_i and Y_j, respectively. Marginal probability density functions are the probability densities for a subset of the data (in this case, looking at the expression profiles of each gene separately). The marginal probability density functions are a measure of how likely it is that an expression value x is a member of the expression profile of its gene X. The probability of x being a member of its expression profile X is low when x is greater or less than would be expected by the rest of the values in X. A high value suggests that there is a potential edge between genes X and Y.

Once a matrix of mutual information scores for each gene pair has been calculated, CLR estimates the likelihood of each pair of scores (a z-score) by comparing them with a background mutual information distribution (MI_i and MI_j). MI_i and MI_j are each just one row of all of the mutual information values for gene$_i$ and gene$_j$, respectively. The idea here is to look at how far away a mutual information score is from the rest of the mutual information scores from that gene. If the score is substantially higher than most of the other scores, there is a good chance that an edge exists. The scores from each gene can be used as a background distribution, because genes, as we have mentioned, tend to depend on only a small number of other genes. Most of the mutual information scores will not be zero due to measurement noise or indirect edges, so these scores can be used as a background noise distribution. Once a z-score for each pair of genes has been calculated, the final step is to calculate the CLR score. It is again calculated for each pair of genes:

$$f(Z_i, Z_j) = \sqrt{Z_i^2 + Z_j^2} \tag{3.8}$$

Z_i and Z_j are the z-scores calculated from the background distribution above. $f(Z_i, Z_j)$ is the joint likelihood measure. This step gives us a single score for every pair of genes that we can compare to the score of each other pair of genes. Finally, we can rank the CLR scores and use the top N scores to build a network. The value of N must be chosen carefully, however, since if the gold standard is unknown then we can't be sure where in the ranking edges become invalid. In practice, the computational analyst will provide this ranking and the experimentalist will test the topmost ranked genes.

3.5.4 Performance on Examples

CLR was tested using the 10-gene DREAM4 network and a DREAM4 100-gene network. Two parameters were tested: the number of bins (from five to 15) and the type of CLR used (there are five types). Each type is a slight variation on the original algorithm. Overall, the version used in [8] outperformed the rest. Due to the large number of parameter combinations tested, only the top five results appear in each graph.

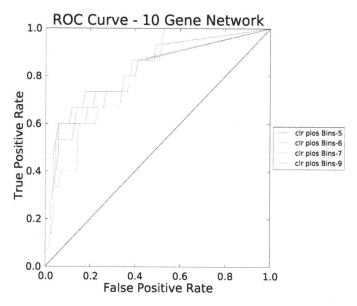

Fig. 3.15 CLR's ROC curve for the DREAM4 10 gene network. (Best AUROC = 0.81 at 6 bins.)

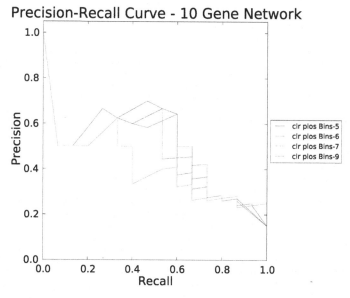

Fig. 3.16 CLR's precision-recall curve for the DREAM4 10 gene network. (Best AUPR = 0.48 at 6 bins)

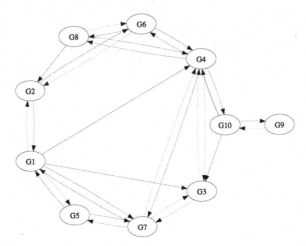

Fig. 3.17 CLR's inferred network for the DREAM4 10 gene network. The top 25% of the ranked edges were used. Solid lines are correct guesses, dotted lines are missed edges, and dashed lines are incorrect guesses. In the top 25% of edges, CLR does well, inferring 10 correct edges, but misses G10's connections.

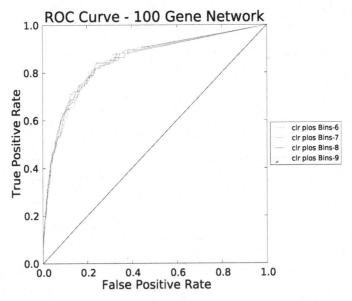

Fig. 3.18 CLR's ROC curve for the DREAM4 10 gene network. (Best AUROC = 0.85 at 8 bins.)

CLR performs very well on the small 10 gene network, and the best performance is achieved using six bins. The precision-recall curve shows that the highest ranked genes from CLR are reliable. The area under the precision-recall curve and the area under the ROC curve are both very high.

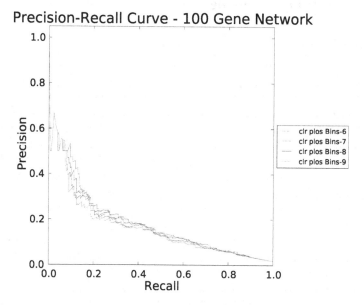

Fig. 3.19 CLR's precision-recall curve for the DREAM4 10 gene network. (Best AUPR = 0.17 at 8 bins)

Performance is also strong on the larger 100 gene network, with eight bins yielding the best performance. Setting the number of bins appropriately is important for CLR's results. In our testing, we found that around 6-8 bins tends to perform the best for networks like these. In practice, CLR was not very sensitive to the value of this parameter on the 100 gene network. The precision-recall and ROC curves look similar for any value of bins above four. CLR was much more sensitive to the parameter on the smaller 10 gene network. In order to estimate the number of bins to use on an unknown network, we ideally would find a known, but similar (in terms of size and data features) network to optimize. If nothing is known, keeping the value small, around five or six, seems to be a good strategy. The precision-recall curve again shows that the highest ranked edges can be trusted to be correct, though the precision falls much faster than in the smaller network. Overall, CLR does a very good job at inferring the network, and provides reliable results across the two datasets.

3.6 Semidefinite Programming

3.6.1 What it does

Convex optimization (or convex programming) is a broad class of optimization algorithms that relies on the assumption that the function being minimized is a convex function that exists in a convex set. A function is convex if a line can be drawn between any two points on the function, and the graph of the function lies below that line. An upward-facing parabola is a simple example. Any line drawn across the parabola has the cup of the parabola below it. Intuitively this method works when the function has a single minimum on the set and there are no deceptive local minima. In other words, the method works when greedy methods work.

Convex optimization generally takes place in the context of constraints. Constraints are used to tell the optimizer what we know about the answer to the problem. As an example, when inferring a gene network using optimization, we may know that certain edges exist, so we can solve the problem with the constraint that those edges are in the answer.

There are many strategies for solving convex optimization problems. The two used here are linear programming and semidefinite programming. A linear program consists of a linear cost function to be minimized, and a series of linear constraints to be considered while optimizing. The space where all of the constraints hold true is the feasible space. A point is chosen in the feasible space, and one of many optimization methods are performed to iteratively move toward the optimal point: minimizing the cost function while meeting the constraints. Semidefinite programming works in a similar way, except that the cost function and constraints are positive semi-definite matrices. Essentially, this relaxes some of the simplifications a linear program has to make, though it becomes slightly more difficult to solve.

Rating: DREAM Fair

3.6.2 The Data

Steady-state data are used to infer the networks using semidefinite programming. Knockdown and overexpression data may be used. Knockout data may perturb the network too much for this approach.

3.6.3 The Strategy

Zavlanos, et al. [31] use semidefinite programming to infer gene regulatory networks. To do this, a network of n genes is modeled as an n dimensional dynamical system. The dynamical system of m steady state experiments can be written as

$$A\tilde{X} + BU = 0, \tag{3.9}$$

where $\tilde{X}$ is an n by m matrix of steady-state expression data, A is the n by n network model, U is an p by m matrix of which genes were perturbed in which experiment, and B is an n by p matrix that indicates which genes are affected by the transcriptional perturbations. Generally, we don't know U, so it is assumed that we can perturb each individual gene, and U is chosen so that BU is a diagonal matrix.

The basic strategy is to first solve a linear programming problem to obtain a matrix A. Then, if the matrix A is unstable, solve a semidefinite programming problem. Continue solving these until A is stable. Stable means that all of the eigenvalues have strictly negative real parts. When this is true, choosing any x for $\dot{x} = Ax$ will eventually converge to the same value, i.e., they are in a stable basin of attraction. The linear program that generates the initial solution for A is

$$\begin{aligned}
\min\ & t \sum_{i,j=1}^{n} w_{ij}|a_{ij}| + (1-t)\varepsilon \\
\text{s.t.}\quad & \|AX + BU\|_1 \leq \varepsilon \\
& \varepsilon > 0 \\
& A \in S,
\end{aligned} \tag{3.10}$$

where S is an n by n matrix of known connections, $\|AX + BU\|_1$ denotes the l_1 norm of matrix $AX + BU$, and ε is the error of the fit between the model and the data in X. The parameter t is a tradeoff parameter between the sparsity of A and the accuracy of the model. The importance of sparsity is increased as t approaches 1, and the importance of minimizing the error is increased as t approaches 0. The weights are chosen by

$$w_{ij} = \frac{\delta}{\delta + |a_{ij}|}, \text{for all } i,j = 1,\ldots,n, \tag{3.11}$$

for some sufficiently small $\delta > 0$. The linear program in (4.14) and the weight update function (3.11) are run until A converges.

Once a final matrix A is obtained from (4.14), a semidefinite program is run iteratively as long as A remains unstable. The goal of this procedure is to find "small" perturbations to A that render it stable while maintaining the necessary constraints. Stability refers to the size of the perturbation required to knock a given matrix of a dynamical system into a new solution. That is, for any starting point of expression values that we choose that is close to the expression values we have, they will settle into the same values over time. However, if the expression values at the starting point are too far away, they will settle to different values. The more stable a matrix, the larger the perturbation required to get it into a new basin of attraction. So, if A is unstable, the idea is to find a nearby matrix A' that is stable, and move to that. The resulting matrix A is then used as our network.

3.6.4 Performance on Examples

Semidefinite programming was tested using the 10-gene DREAM4 network and a larger 20-gene network. For networks larger than 25 genes, the memory requirements become far too large for some parameter combinations. Testing was done over the parameter t at points between 0 and 0.1.

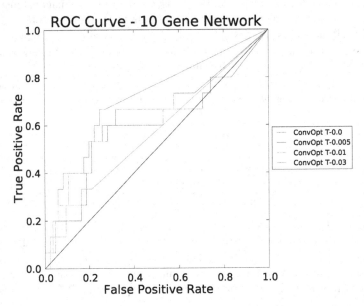

Fig. 3.20 Semidefinite programming's ROC curve for the DREAM4 10 gene network. (Best AU-ROC = 0.70 at $t = 0$)

Semidefinite programming performs well on the small 10 gene network. Notice that when $t \geq 0.06$ the algorithm does no better than random. Thus, it is extremely important to set the parameter t to a good value. In testing, we found that when t is between 0 and 0.05, the algorithm works well. The precision-recall curve shows that for good values of t, the highest ranked genes are reliable. The high area under the precision-recall curve is due to very few guesses being made at $t = 0.04$. Only nine guesses were made, and three of them were correct (leaving six false positives).

The algorithm also performs well on the larger 20 gene network. The high area under the precision-recall curve results from the fact that the algorithm makes fewer guesses at $t = 0.07$. Only 10 guesses were made, and three of them were correct.

One great feature of this algorithm is its ability to directly implement prior knowledge as constraints in the optimization. This allows us to easily create a prior network in another algorithm such as MCZ and prune it using this semidefinite programming algorithm. We'll see more of this later in the Pipelines chapter.

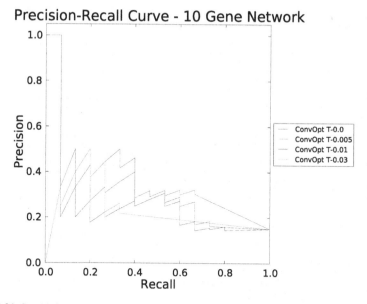

Fig. 3.21 Semidefinite programming's precision-recall curve for the DREAM4 10 gene network. (Best AUPR = 0.63 at $t = 0.04$)

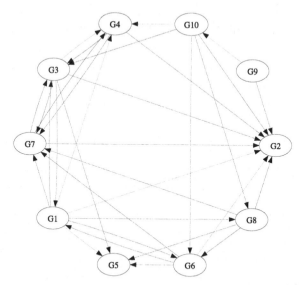

Fig. 3.22 Semidefinite programming's inferred network for the DREAM4 10 gene network. The top 25% of the ranked edges were used. Solid lines are correct guesses, dotted lines are missed edges, and dashed lines are incorrect guesses. In the top 25% of edges, Semidefinite Programming misses some edges, mostly around G1 and G10.

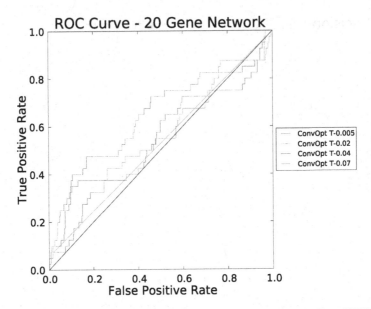

Fig. 3.23 Semidefinite programming's ROC curve for the 20 gene network. (Best AUROC = 0.64 at $t = 0.04$)

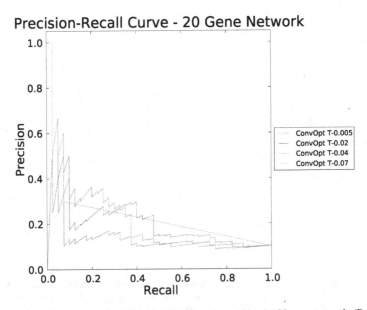

Fig. 3.24 Semidefinite programming's precision-recall curve for the 20 gene network. (Best AUPR = 0.37 at $t = 0.07$)

Chapter 4
Step 3: Using Time-Series Data

4.1 Introduction

Time-series data gives information about the values of genes at a series of consecutive time points. This temporal information can be exploited to infer directionality of edges, or help to infer causal relations between genes. However, adding temporal information also creates a more complex dataset. It adds interdependencies between experiments (time-points) that don't exist in steady-state data, so more care has to be taken in analysis. Three types of algorithms will be presented in this section: mutual information, ordinary differential equations with l_1 regularization, and dynamic Bayesian Networks. Each of these approaches makes different assumptions about the data.

4.2 Time-Delay ARACNE

4.2.1 What It Does

Time-Delay ARACNE is a mutual information based algorithm that works in three steps. First, it detects the point at which each gene begins changing its expression value. The algorithm uses the ratio of change in expression compared to the initial time point to determine when a gene has been substantially induced or repressed. Next, it calculates the mutual information value for each gene pair. If there is a statistical dependency (nonzero mutual information value), then an edge is created between those two nodes. Finally, edges are pruned based on a threshold statistic called the Data Processing Inequality [5], which will be discussed below.

J. M. Lingeman and D. Shasha, *Network Inference in Molecular Biology,*
SpringerBriefs in Electrical and Computer Engineering,
DOI: 10.1007/978-1-4614-3113-8_4, © The Author(s) 2012

Rating: DREAM Fair

4.2.2 The Data

In the examples, Time-Delay ARACNE uses time-series data consisting of 20 closely spaced points.

4.2.3 The Strategy

4.2.3.1 Step 1: Detecting first change in each gene

Time-Delay ARACNE estimates the first time point at which each gene begins to become induced or repressed. It tests whether a gene g is induced by comparing its initial expression value at time 1 to its expression value at time t. The parameter τ is used as a threshold for how much the expression value has to change before the gene is considered induced or repressed. In [32], this value is set to 1.2.

If the ratio between the expression at t and the initial expression value is greater than τ, then the gene is considered induced at time t, as in (4.1).

$$g^+ \text{ if } \tau < \frac{g(t)}{g(1)} \qquad (4.1)$$

Conversely, if the ratio between the expression value at t and the initial expression value is less than than $1/\tau$, then we can consider this gene repressed.

$$g^- \text{ if } \frac{1}{\tau} > \frac{g(t)}{g(1)} \qquad (4.2)$$

If the expression values meets neither of these conditions, it is not considered expressed or repressed at that time point.

There are two benefits to estimating when a gene has begun induction or repression: First, it guides our inference of casual events in the time series. Gene g can influence g' only if g becomes induced or repressed before g'. Second, it allows us to speed up computation by reducing the number of possible edges to examine in steps two and three.

4.2.3.2 Step 2: Detect dependencies between genes at each time point and build a network

Next, Time-Delay ARACNE builds a matrix of mutual information values. Mutual information values are calculated for each pair of genes where gene g begins its expression change at most k time steps before g'. If $k = 3$, for example, then the algorithm calculates the mutual information of gene g at time x with g' for times $x + 1$, $x + 2$, and $x + 3$, for all time points x. For each possible delay value less than or equal to k, there will be a different mutual information value. If the highest such value is greater than a threshold, then an edge is drawn from g to g'. All pairs of genes must have the same delay value. The time point with the highest mutual information value represents the relationship between genes g and g'. If that value is above a certain threshold, then a directed edge is drawn from g to g'. These directed edges will form our network.

The threshold is calculated automatically using stationary block bootstrapping. The data are separated into blocks of expression values, and then sampled with replacement. The randomly chosen blocks are then concatenated to form a new time series. Mutual information values are then calculated for the genes in this new time series. Once this process is iterated a few thousand times, the mean and standard deviation of the mutual information values are taken. The threshold parameter is then defined as $I_0 = \mu + \alpha + \sigma$, where μ is the mean of the mutual information values, α is the statistical significance level (usually 0.05), and σ is the standard deviation. The net effect is that the threshold is a mutual information value well above one standard deviation from the mean mutual information value.

4.2.3.3 Step 3: Trim the network using the Data Processing Inequality (DPI)

Finally, the Data Processing Inequality (DPI) is applied to the network built from step 2. The Data Processing Inequality is a measure that is used to break up three-node cycles. For example, if there is a link going from gene g_a to g_b, from g_b to g_c, and from g_c to g_a, then the link with the lowest mutual information value is removed. However, three-node cycles are allowed to exist if the three mutual information values are all within 15% of each other (this value is from [32]). The idea is to eliminate weak edges that are an artifact of the correlation between the first and third gene in a three-gene chain. For example, if gene g_a has an edge to g_b, and g_b to g_c, then the mutual information value may pick up a weaker edge from gene g_a to gene g_c by mistake. Larger cycles are rarely a problem because there is a lower likelihood of spurious correlation.

Parameter Name	What it does	Default Value
τ	How far a gene's expression value should move from its initial value before it is considered active.	1.2
k	Number of time points to look ahead.	2
Number of bins	The number of bins to use when discretizing the expression ratios.	9

4.2.4 Performance on Examples

Time-Delay ARACNE was tested using the time-series data from the DREAM4 10 and 100 gene networks. Time-Delay ARACNE does not return scores for each edge. Instead, a binary network is returned. Because of this, precision-recall and ROC curves will not be used. Instead, the overall true positive rate (TPR), false positive rate (FPR), and precision will be reported. We will compare the results of using different numbers of bins to discretize the expression ratios.

# Bins	TPR	FPR	Precision
4	0.33	0.10	0.35
6	0.33	0.08	0.41
7	0.33	0.12	0.31
9	0.2	0.02	0.60

Table 4.1 Table of the true positive rate (TPR), false positive rate (FPR), and precision for Time-Delay ARACNE with different numbers of bins on the 10 gene network. Best performance was achieved with seven bins.

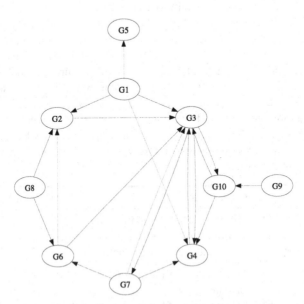

Fig. 4.1 Time-Delay ARACNE's inferred network for the DREAM4 10 gene network. 25% of the ranked edges were used. Solid lines are correct guesses, dotted lines are missed edges, and dashed lines are incorrect guesses. In the 25% of edges selected, Time-Delay ARACNE is able to pick up five edges.

On the small 10 gene network, Time-Delay ARACNE recovers only five of 15 edges, but with 11 false positives using seven bins. So while Time-Delay ARACNE is unable to infer the entire network, the edges that it does infer seem to be fairly accurate. It guessed some false positives, but it still did far better than random chance.

# Bins	TPR	FPR	Precision
4	0.13	0.059	0.038
6	0.13	0.059	0.038
7	0.13	0.059	0.038
9	0.13	0.059	0.038

Table 4.2 Table of the true positive rate (TPR), false positive rate (FPR), and precision for Time-Delay ARACNE with different numbers of bins on the 10 gene network. Time-Delay ARACNE converged to the same network regardless of the number of bins.

However, for the 100 gene network, the results are very different. The number of bins does not seem to influence the result in this case, as each run of the algorithm converged to the same network. Time-Delay ARACNE inferred only 23 true positive edges with 580 false positive guesses. Time-Delay ARACNE performs badly on this DREAM experiment but performs well on its original data set. One reason might be that Time-Delay ARACNE was built for small networks. Another reason is that the simulated data used here may have different characteristic (e.g., noise profile) than the real-world data Time-Delay ARACNE was built for, and thus may not work well with this data.

4.3 Time-Lagged Context Likelihood of Relatedness (tlCLR)

4.3.1 What it Does

Time-Lagged Context Likelihood of Relatedness (tlCLR) [11] [20] is an extension
of the CLR algorithm (previously discussed in Section 3.5) used for inferring networks from time-series data. It is part of the Inferelator 2.0 pipeline that scored
well in the DREAM4 competition [11], along with its counterparts Inferelator and
Median-Corrected Z-Scores.

Rating: DREAM Good

4.3.2 The Data

tlCLR uses both steady-state and time-series data to infer networks. Time-series
data add the ability to infer the directionality of the edges and that is what we focus
on in this section.

4.3.3 The Strategy

tlCLR is a mutual information based inference algorithm that uses ordinary differential equations to model time-series data. The algorithm works in three steps: first,
the temporal changes in expression are modeled as an ordinary differential equation
(ODE). Second, the mutual information is calculated between each pair of genes.
Third, a background correction is applied to filter out less likely connections.

4.3.3.1 Step 1: Modeling time-series data with ordinary differential equations

In order to calculate the mutual information for time-series data, we have to create
a measure of the expression value that captures its change over time. To do that,
tlCLR assumes that the temporal changes in a gene's expression can be modeled
with a linear ordinary differential equation (ODE). The ODE can be simplified using
finite difference approximations, and we can represent the time series as a set of
expression values and response variables:

$$y_i(t_{k+m}) = \tau_i \frac{x_i(t_{k+m}) - x_i(t_k)}{t_{k+m} - t_k} + x_i(t_k), m = 1, 2 \qquad (4.3)$$

where $y_i(t_{k+m})$ is the "response variable" of gene i at time t_{k+m}, where k is the current time and m is the number of time points to look ahead. The response variable is a measure of the expression value's change between time t and time $t + m$. Since the steady-state expression values are not changing over time, the response variable for a steady-state experiment e is just set to its expression value ($y_i(l) = x_i(l)$). $x_i(t_k)$ is the expression value of gene i at time k, and $t(k)$ is the kth time value. τ_i is the inverse of the first order degradation rate for gene i. This value is related to the half-life (in minutes) of the mRNA and must be obtained from literature for each organism. In [20], the first order degradation rate τ_i is set to 10 minutes for all genes. This value is within the range of measured half-life times for $E.Coli$.

After calculating response variables for the time-series data and steady-state data, we end up with a vector of response variables y_i and a vector of corresponding explanatory variables (expression profiles) x_j for each gene:

$$y_i = (y_i(t_2),\ldots,y_i(t_k),y_i(t_3),\ldots,y_i(t_k),y_i(e_1),\ldots,y_i(e_M)), \quad (4.4)$$

$$x_j = (x_j(t_1),\ldots,x_j(t_{k-1}),x_j(t_1),\ldots,x_j(t_{k-2}),x_j,(e_1),\ldots,x_j(e_M)), \quad (4.5)$$

where the first batch of time-series values are for when $m = 1$, and the second for when $m = 2$. The remaining values are the steady-state values, denoted by e. It should be noted that the predictor variables are time-lagged with respect to the response variables, e.g., that predictor variables are lagging m steps behind the response variables. There is no explicit regularization to reduce the number of independent variables that could influence a response variable, but only the top-ranked edges are retained.

tlCLR computes two different sets of mutual information values. The first is called static mutual information, denoted by matrix M^{stat}. Each entry of the matrix is the traditional mutual information value, calculated between the expression profiles of each pair of genes $I(x_i,x_j)$, the same way the mutual information is calculated in ARACNE and CLR. In this case, time-series data are treated as if they were a collection of independent steady-state expression, and the entire vector x_j is used as the expression profile. Note that because $I(x_i,x_j) = I(x_j,x_i)$, the matrix M^{stat} is symmetric. The second set of mutual information values, and tlCLR's main innovation, is dynamic mutual information, denoted as matrix M^{dyn}. This is the mutual information between the response vector y_i at a later time and expression profile x_j for each pair of genes, $I(y_i,x_j)$. Since $I(x_j,y_i) \neq I(x_i,y_j)$, M^{dyn} is an asymmetric matrix, allowing us to infer directionality. What we're looking at is whether gene j's expression value $x_j(t)$ gives us information about gene i's response value $y_i(t+1)$. As in CLR, a high mutual information value implies a likely edge whose target is the response value.

4.3.3.2 Step 3: Ranking the edges

Similar to CLR, a background correction is applied by calculating z-scores for each pair of genes. This allows us to capture information about the degree to which the regulator j determines changes in gene i. Two z-scores are calculated for each pair of genes i and j, determining whether gene j regulates gene i. The first z-score uses only the dynamic mutual information values, while the second uses both dynamic and static mutual information values. The first z-score calculated with respect to the ith row of M^{dyn}, is defined as:

$$z_1(x_i, x_j) = max\left(0, \frac{M_{ij}^{dyn} - \frac{\sum_{j'} M_{i,j'}^{dyn}}{N}}{\sigma_i^{dyn}}\right) \tag{4.6}$$

where σ_i is the standard deviation of the ith row of M^{dyn}. The z-score measures how many standard deviations the dynamic mutual information value between genes i and j is away from the mean of gene i's dynamic mutual information values. The more positive the z-score, the more likely it is that there is some interaction between the two genes. Only positive z-scores are considered. Negative z-scores are set to 0, because they represent absence of positive or negative causality.

Similarly, the second z-score is calculated using both the dynamic and static mutual information values with respect to the jth column of M^{stat}:

$$z_2(x_i, x_j) = max\left(0, \frac{M_{ij}^{dyn} - \frac{\sum_{i'} M_{i',j}^{stat}}{N}}{\sigma_j^{stat}}\right), \tag{4.7}$$

where σ_j^{stat} is the standard deviation of the jth row of M^{stat}. This z-score represents the number of standard deviations the dynamic mutual information value is from the mean of gene x_j's static mutual information values.

These two values for each pair of genes are then combined into a CLR-pseudo z-score:

$$z_{ij}^{tlCLR} = \sqrt{z_1^2 + z_2^2}, \tag{4.8}$$

which is a score that should be proportional to the likelihood that gene j regulates gene i. This ranking is then used to generate a list of the most likely edges, allowing us to build our network from the top ranked edges.

4.3.4 Performance on Examples

tlCLR was tested on the DREAM4 10-gene network and DREAM4 100-gene network datasets, using all steady-state (wildtype, knockout, and knockdown) and time-series data, and then tested using all of the data except knock-down. Overall, it performs well on both datasets. The ROC behavior is consistently better than random behavior. For the ten gene network, the precision-recall curve tells us that the highest ranked edges are likely to be true edges.

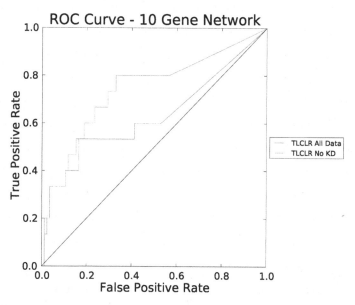

Fig. 4.2 tlCLR's ROC curve for the DREAM4 10 gene network. (AUROC = 0.63 for all data)

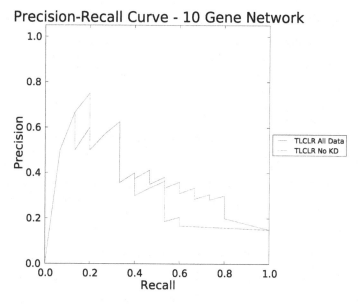

Fig. 4.3 tlCLR's precision-recall curve for the DREAM4 10 gene network. tlCLR performs reasonably well. (AUPR = 0.32 for all data)

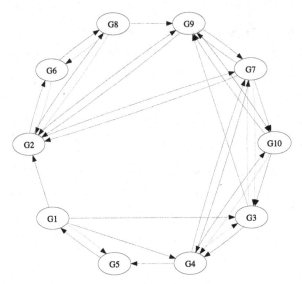

Fig. 4.4 tlCLR's inferred network for the DREAM4 10 gene network. The top 25% of the ranked edges were used. Solid lines are correct guesses, dotted lines are missed edges, and dashed lines are incorrect guesses. In the top 25% of edges tlCLR infers several incorrect edges going to and coming from G2 and is able to recover the relationship between G1 and G5.

Interestingly, the knockdown data actually appears to negatively affect the outcome. This could be because the knockdown data are too noisy.

tlCLR also performs well for the 100 gene network. The ROC curve shows good performance, and the precision-recall curve shows again that the highest ranked edges are accurate. However, this quickly drops off after the first ten or so edges. In the case of the larger network, having extra steady-state experiments from knockdown data improves the results.

Overall, tlCLR is a good algorithm for inferring gene regulatory networks from a combination of time-series and knock-out data. The highest ranked edges are generally reliable. However, in contrast to some of the other algorithms that use time-series data (such as Dynamic Factor Graphs or Inferelator), it does not infer any new information about the dynamics of the network, so this output cannot be used to predict a network's response to new perturbations.

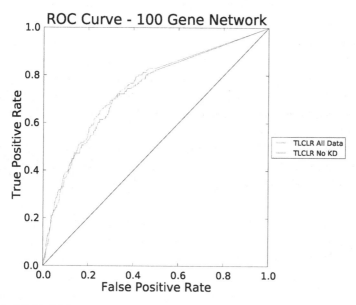

Fig. 4.5 tlCLR's ROC curve for the DREAM4 10 gene network. (AUROC = 0.73 for all data)

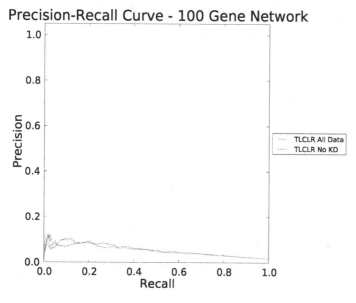

Fig. 4.6 tlCLR's precision-recall curve for the DREAM4 100 gene network. tlCLR again performs reasonably well, with highly ranked edges being accurate. (AUPR = 0.05 for all data)

4.4 Inferelator

4.4.1 What it Does

Inferelator is an inference algorithm based on ordinary differential equations that learn a sparse dynamical model for each gene using data in a time series. The algorithm is part of the DREAM4 Inferelator Pipeline [11] and the DREAM3 Inferelator Pipeline [20].

Dynamical models learn how each gene is changing over time as a function of how its potential regulators are changing. The space of possible potential regulators may be all genes or just the subset of genes that pass some previous test such as Median-Corrected Z-Scores or tlCLR.

Rating: DREAM Good

4.4.2 The Data

Inferelator builds a dynamical model from time-series and steady-state data. In preparation, each experiment should be normalized to have mean 0 and standard deviation 1. The mean of an experiment E is subtracted from each value in E, and then divided by the standard deviation of E.

4.4.3 The Strategy

Inferelator assumes that the changes in a gene's expression value over time are governed by an ordinary differential equation (ODE):

$$\frac{dx_i(t)}{dt} = -\alpha_i x_i + \sum_{j=1}^{p^i} B_{ij} x_j^{r_i}(t), \qquad (4.9)$$

where α_i is the first-order degradation rate of the mRNA of gene i, x_i is the vector of expression values across the time-series for gene i, x^{r_i} is the list of potential regulators for gene i, and p_i are the indexes of the p_i potential regulators for gene i. The list of potential regulators may be all other genes, or come from prior knowledge about the dataset (from experimental data or the output of another algorithm). The first-order degradation rate must be obtained from literature about the organism being studied. In [11], where *E. Coli* was used, this value was set to 10 minutes.

The idea here is that we want to estimate how each gene's expression value changes between time points. It is assumed that each gene i's expression value at time $t+1$ changes as a function of its expression value at time t and the expression values of its regulators at time t. Regression is used to estimate how much each other gene affects gene i (represented by the B matrix). In order to enforce the sparsity of the matrix, a Least Angle Regression (LARS) [7] is used. The basic idea behind LARS is to force all but the strongest relationships between genes to 0. This yields a parsimonious result: each gene ends up having relatively few regulators.

LARS works by minimizing the least-squares error for each gene:

$$E(B) = \sum_{i=1}^{N} E_i(B) \tag{4.10}$$

where

$$E_i(B) = \sum_{r=1}^{R} \left| y_i(r) - \sum_{j=1}^{p_i} B_{ij} x_j^{r_i}(r) \right|^2 , \tag{4.11}$$

where $y_i(r)$ is the rth value of the response variable i. [1]

The quantity $E_i(B)$ in Equation (4.11) is the difference between the actual change of gene i's expression value between times t and $t+1$, and the estimated change of the expression value of gene i at time t based on the expression values and weights of gene i's potential regulators p_i as determined by the model.

By minimizing Equation (4.11) with no further penalties, we may over fit the least squares error estimate, B^{ols}. To combat this, LARS enforces an l_1-norm regularization penalty on the regression coefficients in Equation (4.11), resulting in:

$$\sum_{j=1}^{p^i} |B_{ij}| \leq s_i \sum_{j=1}^{p^i} |B_{ij}^{ols}| , \tag{4.13}$$

where s_i is a shrinkage coefficient (where $s \leq 1$) chosen by ten-fold cross validation. The net effect is that there is a penalty for overly large coefficients on regulators. That tends to give a sparse result, thus helping to avoid over-fitting. The problem can be rewritten in the form:

---

[1] This is the same response variable as defined in the Time-Lagged Context Likelihood of Relatedness algorithm:

$$y_i(t_{k+m}) = \tau_i \frac{x_i(t_{k+m}) - x_i(t_k)}{t_{k+m} - t_k} + x_i(t_k), m = 1, 2 \tag{4.12}$$

where τ_i is equal to the inverse of the first-order degradation rate α of gene i. $x_i(t_k)$ is the expression value of gene i at time t_k, and m is a parameter for how many time points we will look ahead. For steady-state data, the response value y_i is set to the expression value of gene i. The value calculated in Equation (4.12) is a measure of the change between time points t_k and t_{k+m} for gene i's expression value.

$$\min \sum_{r=1}^{R} \left| y_i(r) - \sum_{j=1}^{p_i} B_{ij} x_j^{r_i}(r) \right|^2$$

$$\text{s.t.} \quad \sum_{j=1}^{p^i} \left| B_{ij} \right| \leq s_i \sum_{j=1}^{p^i} \left| B_{ij}^{ols} \right|, \qquad (4.14)$$

The resulting model in B can be used to build the network. Each row i of B is a dynamical model of gene i, where non-zero weights represent connections to other genes. Because it is a dynamical model, it can also be used to predict expression values at future time points. Another interesting application of this is that the model can be used to predict what happens to the network when new perturbations are introduced. For instance, to simulate what happens when a gene is knocked out, we can set that gene's expression value to 0 and then check the response of the network based on the already discovered B.

The dynamical model can also be turned into a list of ranked potential edges. One naive way to do this is to just rank the edges by their value in B. However, this does not necessarily map to a ranking because the weight in the model does not fully represent the explanatory power of gene j as a regulator of i.

We measure the explanatory power of the model for each gene by summing the relative error between each gene's measured response vector and the model-predicted response vector. The edges we keep are the ones having the smallest errors.

4.4.4 Performance on examples

The Inferelator algorithm requires a ranked list of potential regulators that come from another algorithm. Please see Section 5.4 for an example of Inferelator used as part of a pipeline.

Parameter Name	What it does	Default Value
m	Number of time points to look ahead.	2

4.5 Dynamic Factor Graphs (DFG)

4.5.1 What it does

The Dynamic Factor Graphs (DFG) algorithm used here [19] is an ordinary differential equation based method that separates the problem of experimental noise from network inference itself. The method models experimental noise as a fitted Gaussian and then tries to infer networks based on an assumed underlying idealized gene expression. Predictions about the noisy dataset are made assuming a fitted Gaussian noise model. The procedure uses bootstrapping to calculate p-values for each weight in the model, and only those weights with p-values below the cutoff value are used in the final network.

Rating: DREAM Fair

4.5.2 The Data

Typically, biological experiments are repeated a few times (usually less than five times) to compensate for both biological noise and instrument error. Many approaches use the mean values of such collections of replicates. DFG uses each replicate independently.

4.5.3 The Strategy

DFG uses a graphical model to create a relationship between genes and time points, and the algorithm works by iteratively performing inference and learning steps. The algorithm attempts to model both the unknown idealized vector of Z values (one entry per gene) and the family of functions f that models the relationship $Z(t) = f(Z(t-1))$. The dynamical model f is represented by an n by m matrix F, where n is equal to the number of genes, and m is equal to the number of transcription factors That is, F models the dependency of each gene g at time t on a small set of genes (possibly including g) at time $t-1$. The hope is to find an f that is time-independent. There are two constraints on Z given a current guess f: the Z values must be close to the mean of the observed Y values with some allowance for Gaussian noise while also satisfying $Z(t) = f(Z(t-1))$. The algorithm is iterative, adjusting both Z and f as it goes.

The dynamical model is characterized by the following equation:

$$\frac{\tau}{t_{k+1} - t_k} (z_i(t_{k+1}) - z_i(t_k)) + z_i(t_k) = \sum_{j=1}^{N^i} F_{ij} z_j(t_k) + \beta_{i0} + \eta_i(t_k) \tag{4.15}$$

where τ is a parameter that determines the importance of the amount of time between two time points, β is a bias term, and η is a Gaussian error term with zero mean and fixed covariance. The model weights and latent variables are initialized randomly.

A second model, the observational model h, is used between $Y(t)$ and $Z(t)$. This model is essentially an n by n identity matrix with a Gaussian error term.

The values observed at time t, denoted Y(t), are assumed to be the same as the idealized values, denoted Z(t), plus a Gaussian error term.

$$y_i(t) = z_i(t) + \varepsilon_i(t) \tag{4.16}$$

4.5.3.1 Learning the Dynamical Model

Learning the dynamical model f (and associated matrix F) uses uses LARS (Least-Angle Regression) [7] to minimize the quadratic error of only the dynamical model. The idea behind LARS is to produce simple, parsimonious models by iteratively selecting explanatory variables based on their relationship to the residual and other selected explanatory variables. Other optimization methods such as conjugate gradient or Elastic Nets could be used instead of LARS. An l_1 regularization is implemented by LASSO [29] to enforce sparsity of the model f. LASSO aggressively shrinks the weights in F, forcing all but the strongest weights to 0.

4.5.3.2 Bootstrapping

The above iteration between modeling f and reducing the difference between observed and hidden variables is run 100 times, but because it relies on a random starting point for the weights and latent variables, slightly different answers may be obtained from different runs. This issue is handled in two steps. First, the algorithm is repeated 20 times, each returning a new model matrix $F^{*(k)}$. A new model matrix F^* is created by averaging the 20 separate models. Next, a bootstrapping method is used to obtain a p-value for each weight. Each of the 20 models are randomly permuted, and then averaged again. The process is repeated 1,000 times in order to generate a distribution we can use to calculate the statistical likelihood of a model weight randomly occurring. We then calculate a p-value for each weight F_{ij}^*, and if the p-value is below 0.001, we accept it as a valid edge. Taking all of the edges with p-values below this cutoff value yields the inferred network.

We can also obtain a ranking of the edges by giving each of them a z-score:

$$z_{ij} = \frac{F_{ij}}{\beta_{i0} + \sum_{j=1}^{N} F_{ij}} \tag{4.17}$$

where F_{ij} are weights in the model matrix F, and β_{i0} is the corresponding bias value.

Parameter Name	What it does	Default Value
τ	Importance of time between time points	3
Number of models	How many models to build for averaging/boot strapping	20
λ_w	Importance of latent variables vs. observed variables in the inference step.	0.01
η_z	Weight on the Gaussian error term	0.1
p-value cutoff	The cutoff p-value where a weight is considered significant	0.001

4.5.4 Performance on examples

4.5.4.1 The small network example

For this example we use the time-series data from the small 10-gene DREAM4 network. DFG is sensitive to its hyper-parameters, so a small grid search is done to find a good set. The receiver-operator curve figure 4.7 and table 4.3 shows exactly how sensitive it is to the hyperparameter selection, especially for small networks.

η_z	λ_w	τ	AUROC	AUPR
0.1	0.01	3.5	0.38	0.11
0.001	0.001	3	0.42	0.12
0.0001	0.001	1	0.60	0.18
0.0001	0.2	1	0.60	0.18
0.0001	0.001	3	0.48	0.14
0.0001	0.2	3	0.48	0.14
0.1	0.001	3	0.41	0.12
0.1	0.001	1	0.35	0.10
0.1	0.2	1	0.35	0.10
0.1	0.2	3	0.41	0.12

Table 4.3 Table of areas under the receiver-operator curve (AUROC) and the precision-recall curve (AUPR) for Dynamic Factor Graphs for the small 10 gene dataset at different parameter combinations found in a very coarse grid search. Different parameter combinations can greatly affect the results.

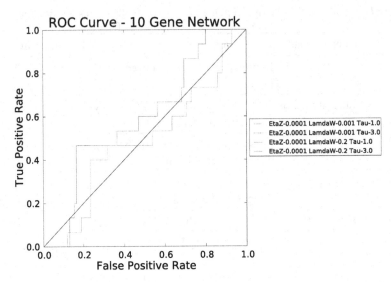

Fig. 4.7 DFG's ROC curves for the DREAM4 10 gene network. The results where η_z and λ_w are equal yielded the exact same network for the two values of τ, so their lines overlap perfectly. (Best AUROC = 0.60)

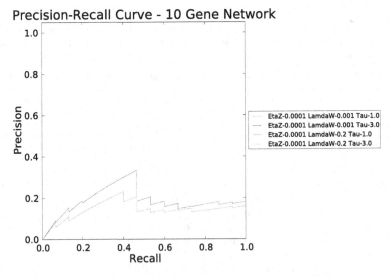

Fig. 4.8 DFG's PR curves for the DREAM4 10 gene network. DFG struggles to maintain good precision, even with its highest ranked guesses. The results where η_z and λ_w are equal yielded the exact same network for the two values of τ, so their lines overlap perfectly. (Best PR = 0.18)

We can see that on the small network, DFG performs best when $\eta_z = 0.0001$ and $\tau = 1$. It doesn't seem to matter whether λ_w is set to 0.001 or 0.2 in this case. These results underscore the importance of finding a good set of parameters.

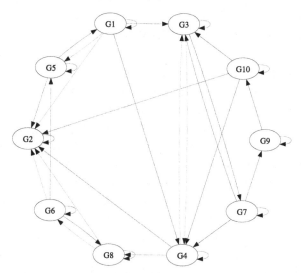

Fig. 4.9 DFG's inferred network for the DREAM4 10 gene network. The top 25% of the ranked edges were used. Solid lines are correct guesses, dotted lines are missed edges, and dashed lines are incorrect guesses. In the top 25% of edges, DFG is able to infer the connections to G2, but misses many of the edges going to G4.

4.5.4.2 The large network example

η_z	λ_w	τ	AUROC	AUPR
0.1	0.01	3.5	0.58	0.02
0.001	0.001	3	0.58	0.02
0.0001	0.001	1	0.60	0.03
0.0001	0.2	1	0.58	0.02
0.0001	0.001	3	0.60	0.03
0.0001	0.2	3	0.58	0.02
0.1	0.001	3	0.58	0.02
0.1	0.001	1	0.63	0.03
0.1	0.2	1	0.58	0.02
0.1	0.2	3	0.63	0.03

Table 4.4 Table of areas under the receiver-operator curve (AUROC) and the precision-recall curve (AUPR) for Dynamic Factor Graphs for the large 100 gene dataset at different parameter combinations. Different parameter combinations can greatly affect the results.

So what is happening here? The ROC curves are decent, but the Precision-Recall curve is terrible. As recall increases, the precision is remains low. This means that the algorithm is making a lot of wrong guesses for each correct guess. DFG does not perform well on these datasets.

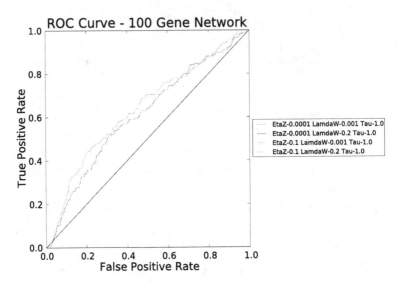

Fig. 4.10 DFG's ROC curves for the DREAM4 100 gene network. The results where η_z is equal yielded the exact same network for the two values of λ_w and τ, so their lines overlap perfectly. (Best AUROC = 0.63)

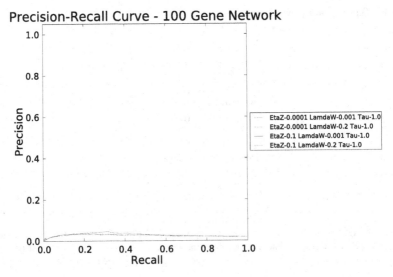

Fig. 4.11 DFG's PR curves for the DREAM4 100 gene network. Even DFG's highest ranked guesses do not do well. The results where η_z is equal yielded the exact same network for the two values of λ_w and τ, so their lines overlap perfectly. (Best PR = 0.03)

4.5.4.3 Overall Performance

While DFG did not perform very well on either the small or large network, it is worth noting that it was using only the time-series data. This data does not have the kind of rich information that a direct genetic perturbation like a knock-out can give.

One good practical feature of DFG is that it yields a sparse network instead of a ranked list (though it can do that too). This may be beneficial in some situations where one wouldn't know how to select a cutoff value for a list of ranked edges.

It is extremely important to find a good combination of hyperparameters for DFG. However, this can be difficult without a full or partial gold standard for comparison. One option for finding parameters for a network that does not have a partial gold standard is to use the dynamics that DFG calculates to predict the last one or two time points. Based on how well these predictions fit, we can have some confidence that the parameters are good.

4.6 Bayesian Network Inference with Java Objects (BANJO)

4.6.1 What it does

Bayesian Network Inference with Java Objects (BANJO) [30] infers edges networks that can be modeled as a first-order Markov process. This means that each gene's expression can be predicted by some combination of the expression values of its parents and itself at the previous time point. No assumptions are made about linearity or non-linearity. The Markov processes work on discretized values, so the algorithm is sensitive to the number of bins the expression values are discretized into.

The basic approach of BANJO is to search through all possible networks, looking for the network with the best score. Of course, for any non-trivial dataset, we can't actually look at all possible network combinations, so various search strategies are used: greedy, simulated annealing, and genetic algorithms. The scoring is done using one of two metrics: Bayesian Dirichlet Equivalence (BDE) [13] or Bayesian Information Criterion (BIC) [28]. The greedy search algorithm and the Bayesian Dirichlet equivalence scoring method were found to work best for gene networks in [30].

In [30], all three search algorithms found the best network for the test datasets (which are different from DREAM), but the greedy algorithm found it substantially faster than simulated annealing or genetic algorithms. Both Bayesian Dirichlet Equivalence and Bayesian Information Criterion worked well as scoring measures for large datasets (hundreds of genes and thousands of time points), but for small datasets Bayesian Dirichlet equivalence was able to score the networks more accurately than Bayesian Information Criterion. BANJO also uses an influence score to guess the directionality of the edges.

Rating: DREAM Fair

4.6.2 The Data

BANJO uses time-series data to infer networks. Since Bayesian approaches to gene network inference typically suffer from lack of data, BANJO attempts to augment the time-series dataset using linear interpolation. It was found in [30] that interpolating one time point between observed time points helped the scoring methods BANJO uses. BANJO is capable (and in fact, seems to work best) when the dataset is huge: hundreds of genes and thousands of time points.

4.6.3 The Strategy

Based on previous uses of BANJO, we'll use as a default: greedy search algorithm with Bayesian Dirichlet Equivalence (BDE) scoring. The greedy search algorithm works by first selecting a random network as its starting point. Call that its current network. The algorithm then evaluates all neighbors to that network according to its BDE score, where a neighbor is the current network with one edge added or removed. The neighboring network with the best score is then declared current, the BDE scores for all of its neighbors calculated, and the network with the highest BDE is selected. This process continues until none of the neighbors have a BDE score higher than the currently selected network. Because greedy algorithms tend to get stuck in local maxima, random restarts are required. This means that the greedy search starts all over again from a new random network. The authors, Yu, et al. perform 100 random restarts. Once the highest ranking network has been selected, an influence score is generated for each edge. This is done in order to predict the sign of the network's edges.

The data are discretized in to three bins prior to any calculations. This is to help simplify the problem because of the relatively small amount of data available. Yu, et al. [30] found through experiment that three bins seemed to be the optimal trade-off between accuracy and the amount of data required. The configuration on our website can be used to change the number of bins.

The BDE is calculated by solving for the log of the marginal likelihood $P(D|G)$ where D is the data and G is the network graph. To do this, we integrate over all possible parameter assignments Θ:

$$logP(D|G) = log \int_{\Theta} P(D|G,\Theta)P(\Theta|G)d\Theta \qquad (4.18)$$

There are two intuitive reasons that make it attractive for scoring gene networks. The first is that the scores are better when a parent is better at predicting a child. The second is that this score penalizes complexity: the more parents that a child has, the lower the score. [30]

The influence score proposed by [30] is based on comparing the expression values of a gene to that of its children. If it tends to be high when its children are high, and low when its children are low, it is an activator. If it tends to be high when its children are low, and low when its children are high, then it is a repressor. The inferred network is built from these influence scores.[2]

---

[2] First, a table of cumulative density function (CDF) values is built from Θ_{ijk} where Θ_{ijk} is the probability that gene X_i is in expression state k when its parents are in expression state configuration j. The expression state configuration is the combination of discretized expression values for all of the parents of some gene. For example, if gene g has three parents, and the expression values have been discretized into $k = 3$ states, then a possible configuration j is when parent $p1$ is in state 0, $p2$ is in state 1, and $p3$ is in 1. For three parents when $k = 3$, there are 27 possible combinations, so j is a number between 0 and 26. The CDF value c_{ijk} is the probability that a child node X_{ijk} is in state k or lower when its parents are in configuration j.

Parameter Name	What it does	Default Value
Search type	Valid options are greedy, simulated annealing, and genetic algorithm. This selects the algorithm to use to move through the network search space.	greedy
Scoring measure	Valid values are the Bayesian Dirichlet Equivalence (BDE) or the Bayesian Information Criterion (BIC). BDE seems to work best with gene networks.	BDE
k	The number of bins to split the data into. The algorithm is very sensitive to this parameter.	3
w	How many points should be interpolated between observed data points?	1

4.6.4 Performance on Examples

BANJO was tested using the time-series data from the DREAM4 10 and 100 gene networks. Because of the relatively small number of time points involved in these datasets, BANJO was unable to calculate non-zero "influence scores" for each edge. This made it impossible to rank edges in relation to each other: only a binary classification is available. Instead, the overall true positive rate (TPR) and false positive rate (FPR) will be reported, as well as precision.

Different runs of BANJO were compared by manipulating the number of bins used to discretize expression ratios (from two to nine bins). In general this is the most important parameter to choose properly, though BANJO has many other parameters and options to experiment with.

$$c_{ijk} = \sum_{k'=0}^{k} \Theta_{ijk'} \tag{4.19}$$

The intuition here is that if the parent is an activator, the CDF value should shift in the positive direction as the value of the parent increases. The reason is that if the parent is an activator, then the child gene's expression value should rise and fall with the parent. However, if the parent is a repressor, than the CDF value should shift in the negative direction as the value of the parent increases.

To figure out the direction of an edge in the case of a child gene having multiple parents, a voting system is used. For each parent gene p of some gene g, to determine whether p is active or repressive, all other parents of g are held at fixed states. The effect on gene g is then recorded as a vote. When this is complete, one of the frozen parents will have its state shifted by one, and the algorithm recalculates how p affects the child g. This continues until all combinations of the frozen parents genes have been tested, and the votes from each combination are tallied. If the votes are all positive or positive and neutral, the gene is marked as an activator of the child. If the votes are all negative or negative and neutral, the gene is marked a repressor of the child. If the votes are a combination of positive and negative, then no inference on the directionality of the edge can be made and the influence score is set to 0.

# Bins	TPR	FPR	Precision
2	0.466	0.388	0.175
3	0.466	0.388	0.175
4	0.466	0.388	0.175
5	0.4	0.380	0.157
6	0.466	0.388	0.175
7	0.6	0.466	0.225
8	0.533	0.409	0.181
9	0.466	0.388	0.175

Table 4.5 Table of the true positive rate (TPR), false positive rate (FPR), and precision for BANJO with different numbers of bins on the 10 gene network. Best performance was achieved with 7 bins.

BANJO performs best on the 10 gene network when using seven bins. The performance is significantly affected by the number of bins. With seven bins, BANJO was able to recover 9 of 15 edges, but guessed on 40 of those edges. This is not a very good result.

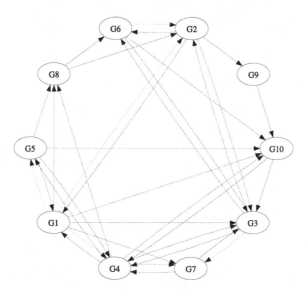

Fig. 4.12 BANJO's inferred network for the DREAM4 10 gene network. 25% of the edges were used. Solid lines are correct guesses, dotted lines are missed edges, and dashed lines are incorrect guesses. BANJO is able to infer only a few edges.

Unfortunately BANJO performs even worse on a larger network. Only a handful of edges are correctly inferred with many false positives. This makes sense, as the

# Bins	TPR	FPR	Precision
2	0.03	0.06	0.01
3	0.10	0.06	0.02
4	0.09	0.06	0.02
5	0.05	0.05	0.01
6	0.05	0.06	0.01
7	0.07	0.06	0.02
8	0.07	0.06	0.02
9	0.08	0.05	0.02

Table 4.6 Table of the true positive rate (TPR), false positive rate (FPR), and precision for BANJO with different numbers of bins on the 100 gene network. Best performance was achieved with 3 bins.

number of time points to number of genes ratio falls sharply (it is now trying to infer the relationships among 100 genes from 20 time points instead of the relationships among 10 genes). Bayesian methods are most commonly used for datasets that have many densely packed time points (e.g. neuronal data). In short, BANJO is not an algorithm to be used for inferring gene regulatory networks with a small number of time points. BANJO performs badly on this DREAM experiment but performs well on its original data set. One reason might be that this data does not have enough time points, or the time points are too far apart to work with BANJO.

Chapter 5
Step 4: Pipelines

5.1 Consensus Step: Combining Results of Different Approaches

Various techniques for gene network inference learn from different types of data, have different theoretical approaches, and use different types of statistics. For example, an algorithm using a Bayesian approach may be extracting different information from the data than one using a regression approach. The idea behind the consensus step is to combine the inferential abilities of different methods to arrive at a consensus network. There are many different ways to combine the inferred networks. We will illustrate these ideas in the example pipelines that follow.

5.2 Creating Pipelines and Ensemble Networks

Multiple algorithms can be strung together to combine their results in some way. Interesting combinations can arise from the type of data used for inference (e.g., steady-state vs. time-series), or different theoretical approaches (e.g., mutual information vs. regression).

A pipeline is a sequence of algorithms in which the output of one algorithm feeds into the input of the next algorithm in the sequence. For example, we can use the output of a perturbation algorithm to give a prior weighting of edges, thus giving the next algorithm a good starting point. The first two pipelines in this section take this form.

Ensemble networks are meta-algorithms for combining independent results from different algorithms. For example, we can create an ensemble network from a weighted sum of the networks by finding the weights that optimize the results on data with a gold standard, then use those same weights to create an ensemble from data without a gold standard. The second set of algorithms in this section are ensembles.

J. M. Lingeman and D. Shasha, *Network Inference in Molecular Biology*,
SpringerBriefs in Electrical and Computer Engineering,
DOI: 10.1007/978-1-4614-3113-8_5, © The Author(s) 2012

5.3 Pipeline 1: Steady State algorithm + Dynamic Factor Graphs

5.3.1 What it does

Steady-state algorithm such as Context Likelihood of Relatedness (CLR), Median Corrected Z-Scores (MCZ), and GENIE3 all produce nicely ranked lists of edges for an algorithm like Dynamic Factor Graphs (DFG) to use. DFG uses time-series data to create a sparse network that best explains the given data. The idea behind this pipeline is to use the top edges from the steady-state algorithm's ranked output as a prior for DFG, giving it both a better starting point than a random initialization and restricting the list of possible edges it can choose from.

5.3.2 The Strategy

DFG can use this prior in three ways: 1) we can restrict the possible edges that DFG is allowed to connect, 2) we can boost (or reduce) the initial weights of edges, or 3) do both. Option 1 uses DFG as a pruning algorithm: given a list of ranked edges which are likely to include all of the actual edges, remove the spurious edges. Option 2 uses DFG in its normal form: given these starting weights and data, build a network. Option 3 consists of initializing DFG with both boosted initial weights and a restricted set of possible connections.

There are several sets of parameters to test in this case. The first are the parameters for each of the algorithms. We'll use the set of parameters that yielded the best performance for each algorithm when it is used alone. Next, we have to decide how many of the top ranked edges to use from the steady-state data. And finally, we have to choose which of the three options for using the steady-state data give the best results.

Rating: DREAM Good

5.3.3 Performance on examples

A combinatorial approach was used to test this pipeline. The top N% of edges were taken from each steady state algorithm and combined with DFG, using the steady state algorithm to both initialize weights and restrict possible connections (option 3). This method of combining the steady-state algorithms with DFG was chosen due to consistently better performance than options 1 or 2 in testing. The percentage of edges to use from the steady-state algorithm was done in increments of 10 from 0 to 100.

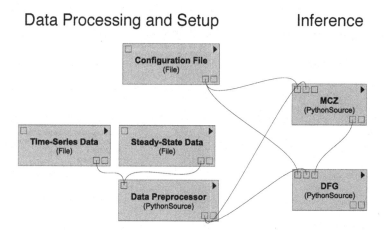

Fig. 5.1 The basic workflow diagram for the MCZ-DFG pipeline algorithm. MCZ was chosen as an example, but it could be replaced by any of the other steady-state algorithms.

Figure 5.2 shows the top 4 results on the small 10 gene network. Two of the top 4 results came from using the steady-state algorithms by themselves, and are shown for comparison (top AUROC was from MCZ at 0.90). The other two top results are MCZ+DFG, using the top 20% and 30% of edges with AUROC of 0.82 and 0.84, respectively. Both of these performed better than DFG alone, which had an AUROC of 0.58 (not shown).

Figure 5.4 shows the top 4 results on the larger 100 gene network. These results are similar to the ones for the 10 gene network. Two of the top 4 results come from using the steady-state algorithms by themselves (MCZ and CLR), and the other three are pipeline results. When using 10% of the edges, the AUROC is 0.86, at 20% the AUROC is 0.84, and at 30% it is 0.82. DFG alone has an AUROC of 0.57 (not shown).

Overall, combining the steady-state data with DFG greatly improves the results of the algorithm. The top 20-30% of the edges from the steady-state algorithm should be used with DFG to maximize performance. These results are very promising. We lose some of the precision of the topology that the steady-state algorithms give us, but we gain the dynamical information from DFG. We can use this dynamical information to predict how the network will respond to new perturbations. The reason is that DFG gives a quantitative prediction of the effect of one gene on other genes in the form of a differential equation. The output of a steady state network shows connections only – there is no quantitative prediction.

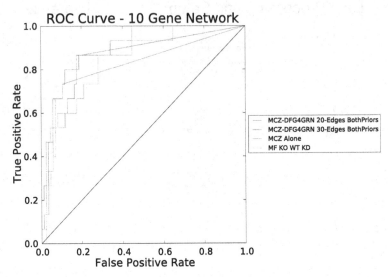

Fig. 5.2 The top 4 results from testing pipeline 1 on the DREAM4 10 gene network. Two of the top 4 results are the steady state algorithms used, and the other two use the top 20% and 30% of the edges from MCZ to both initialize DFG's weights and restrict possible connections. The top AUROC was MCZ alone (0.90). The pipelines had AUROC scores of 0.82 (when using 20% of the edges) and 0.84 (when using 30% of the edges). Combining the results of MCZ with DFG greatly improves the results of DFG.

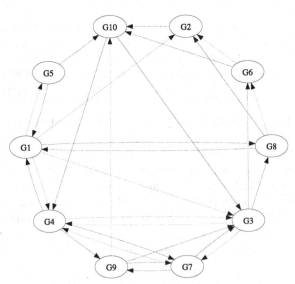

Fig. 5.3 Pipeline 1's inferred network for the DREAM4 10 gene network. The top 25% of the ranked edges were used. Solid lines are correct guesses, dotted lines are missed edges, and dashed lines are incorrect guesses. In the top 25% of edges, Pipeline 1 is able to infer many of the edges, including the edges from G1 that many other algorithms had missed.

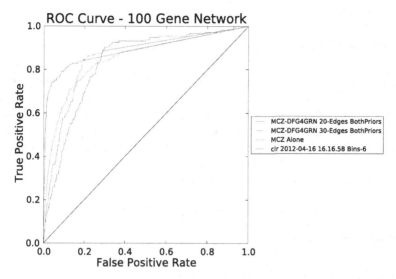

Fig. 5.4 The top 4 results from testing pipeline 1 on the DREAM4 100 gene network. Two of the steady-state algorithms did extremely well on this network (MCZ and CLR), as well as three of the tested pipeline combinations. The top AUROC was MCZ alone (0.90). The pipeline did the best when MCZ was used with DFG, using the top 10, 20, and 30% of edges from MCZ (AUROCs of 0.86, 0.84, and 0.82, respectively). As with the results from the small network, combining DFG with MCZ greatly improved the results from DFG.

5.4 Pipeline 2: Inferelator 2.0

5.4.1 What it does

The Inferelator 2.0 pipeline [11] was a top contender in the DREAM4 challenge. Three algorithms previously discussed (Median-Corrected Z-Scores (MCZ), Time-Lagged Context Likelihood of Relatedness (tlCLR), and Inferelator) are combined using heuristic and resampling techniques. The idea is to create a pipeline that is able to combine information from a simple statistical approach (MCZ) and mutual information (tlCLR), and then use that information to help infer the dynamics of the time-series (Inferelator).

Rating: DREAM Best

5.4.2 The Strategy

Three techniques are used to create a consensus network. (i) The results from tlCLR are used as a feature selection step in the Inferelator algorithm. For each gene, the top P regulators (as predicted by tlCLR) are used as potential regulators for those genes in Inferelator. All other potential edges are set to false. (ii) All three algorithms produce a Z-Score for each edge, and these scores can be combined. (iii) A resampling approach is used to help reduce noise from the dataset.

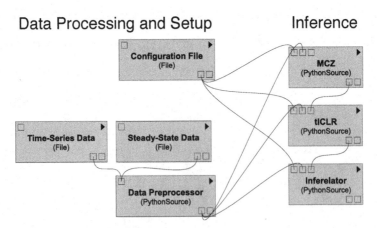

Fig. 5.5 The basic workflow diagram for the Inferelator 2.0 algorithm. Each inference algorithm uses the output from the previous algorithm to help with inference.

Resampling is a technique that replaces each edge value by its non-parametric average. A new dataset is built by randomly sampling from the existing dataset of experiments with replacement, until a dataset of the same size is constructed. A network inference algorithm is then run on the sampled dataset. This procedure is repeated many times to ensure good coverage of the original data. The justification for this approach is that sampling provides a way of looking at what happens when certain experiments drop out of the dataset (does an unrelated edge disappear?), and when certain experiments are duplicated (are some edges stronger?). By averaging the generated networks across these datasets for a given algorithm, we hope to find a network closer to one that we would have found after doing many more experiments.

MCZ is run independently of the rest of algorithms, and results in a matrix of Z-Scores for each edge Z^{mcz}. From the discussion of tlCLR and the Inferelator, recall the definitions of the X design matrix and Y response matrix, both of size N x R. Resampling is done by choosing a random integer between 1 and R, taking the corresponding column from X and Y, and placing it into the new matrices X^* and Y^*, respectively. The tlCLR-Inferelator algorithm is run using X^* and Y^*, computing three matrices: z-scores from tlCLR (Z^{tlCLR}), z-scores from Inferelator (Z^{Inf}), and the dynamical parameters from Inferelator (β). The z-scores of each of these algorithms are then combined with:

$$z_{i,j}^{combined} = \sqrt{\sqrt{(z_{i,j}^{Inf})^2 + (z_{i,j}^{tlCLR})^2} + (z^{mcz})^2}. \tag{5.1}$$

This procedure is repeated B times, where B is set to 200 by default. Finally, the z-scores for each edge i, j from the B samples are ordered and the median value is calculated. The end result is a matrix Z^{median} that contains the median z-score for each edge across the resampling.

5.4.3 Performance on examples

The Inferelator 2.0 Pipeline was tested using the 10 gene and 100 gene networks from the DREAM4 competition. On the 10 and 100 gene networks (Figures 5.6, 5.7), the Inferelator 2.0 Pipeline performs well, but not as well as either of its component steady-state algorithms. Is it still worthwhile to use Inferelator? Like the DFG algorithm in the previous pipeline, Inferelator enables us to discover the dynamics of the network. By first uncovering the topology of the network with MCZ and tlCLR, and then inferring the dynamics based on those topologies, we can make predictions about future time points or future experiments. In short, we trade some accuracy of the topology for new information about the dynamics between the genes.

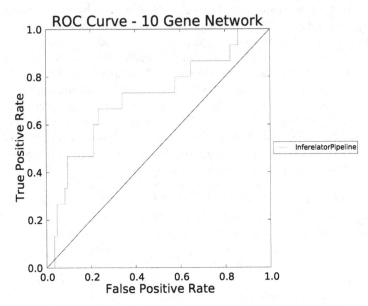

Fig. 5.6 Receiver operating characteristic curves (ROC) of the Inferelator 2.0 Pipeline for the DREAM4 10 gene network. (AUROC = 0.74)

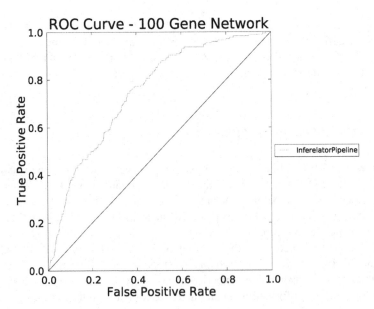

Fig. 5.7 Receiver operating characteristic curves (ROC) of the Inferelator 2.0 Pipeline for the DREAM4 100 gene network. (AUROC = 0.70)

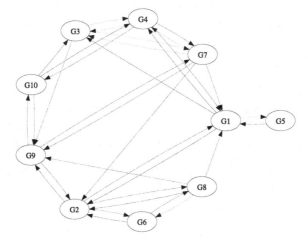

Fig. 5.8 Pipeline 2's inferred network for the DREAM4 10 gene network. The top 25% of the ranked edges were used. Solid lines are correct guesses, dotted lines are missed edges, and dashed lines are incorrect guesses. In the top 25% of edges, Pipeline 2 is able to infer the relationship between G3, G4, and G7, but misses most of the rest of the network.

5.5 Ensemble 1: Voting

5.5.1 What it does

After each network inference algorithm generates a ranked list of potential edges, a combining technique will use the top ranked edges as "votes" for that edge. For example, if in a 4 gene network 4 out of 6 algorithms of different types infer a highly ranked edge between genes 1 and 3, that edge receives 4 votes. The edges are then ranked by the number of votes, and the top *n* edges are selected. Using this method, each algorithm gets an equal vote in the final network. The idea is that some algorithms will miss some edges, but a plurality will pick up on the correct ones.

Rating: DREAM Best

5.5.2 The Strategy

There are many different ways to set up a voting system. One easy way to to have the amount of the vote be a function of its rank in a list of edges. The highest ranked edge will have vote of 1, and that continues along some function that approaches 0.

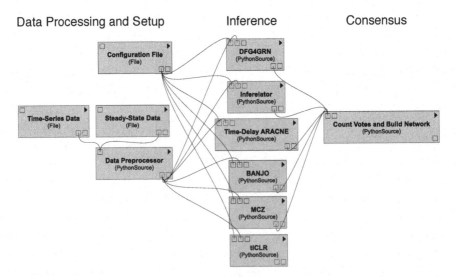

Fig. 5.9 The basic workflow diagram for the voting algorithm described in Section 5.5. The inference algorithms listed could be any number and combination of algorithms that the available data permit.

The reason for this approach is that the top ranked edges from each network are generally the most accurate, so we want to give them a higher score. Most of the time, rankings become inaccurate after only a small number of edges, so we want the drop-off to be severe. To achieve this, the score of each edge e is calculated with:

$$\text{score} = \log(i+1) + \log(n^2) \qquad (5.2)$$

where i is the rank of edge e and n is the total number of genes. These scores are calculated for each edge in each algorithm's output network. Negative scores are changed to 0. This function is used because it will give highly ranked edges a large score, and quickly drop off. For larger networks, the drop off will be slower. The scores of each edge are summed, and these sums are used as edge weights in the final voting network.

5.5.3 Performance on examples

We tested the voting ensemble using several steady-state and time-series algorithms on the five 10-gene and five 100-gene DREAM4 networks. Each of these datasets contains knockout, knockdown, wildtype, time-series, and multifactorial data. The algorithms used in the test were Dynamic Factor Graphs (DFG), Gene Network Inference with Ensemble of trees (GENIE3), Median-Corrected Z-Scores (MCZ), Time-Lagged Context Likelihood of Relatedness (tlCLR), Context Likelihood of Relatedness (CLR), and Network Identification by Multiple Regression (NIR). Each algorithm was given an equal vote. The MCZ algorithm was used as the baseline to compare against, as it was consistently the best performing lone algorithm.

Voting performed roughly on par with MCZ on the 10-gene networks, losing substantially to MCZ on network D but besting it in network E (Figure 5.10). This is a surprising result, given that the algorithms are all given equal weight and most of the algorithms did substantially worse than MCZ. The reason is that this approach relies on heavily weighting the highest ranked edges from each algorithm. If the top edges from each algorithm are correct, and each algorithm is ranking a slightly different set of edges as high, then the overall highest ranked list of edges should be accurate. However, voting is unable to best MCZ on these small networks. What happens if we test this on the larger 100 gene networks?

On the 100 gene networks the results are similar (Figure 5.12). On 3 of the 5 networks, the results were more or less tied, with voting being roughly equal to MCZ. However, on the two networks where MCZ did not do very well, C and E, voting outperformed it. These results suggest that a consensus vote may not help much when algorithms are already able to infer the network well, however, in situations where an algorithm that MCZ seems to falter, other algorithms can step in and correct its mistakes.

Receiver Operating Characteristic (ROC) Curve (10 Genes)

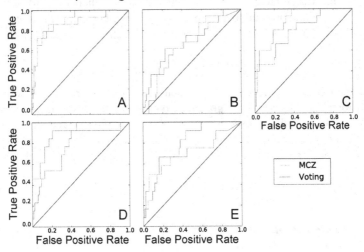

Fig. 5.10 Receiver operating characteristic curves (ROC) of the voting algorithm and median-corrected z-scores (MCZ) for the DREAM4 10 gene networks A-E. Graphs A-C show roughly equal performance between MCZ and Voting. Voting did not greatly outperform MCZ. Graph D shows a case where MCZ outperforms voting (AUROC for MCZ = 0.90, voting = 0.77). Graph E shows a case where voting outperforms MCZ (AUROC for MCZ = 0.71, voting = 0.79).

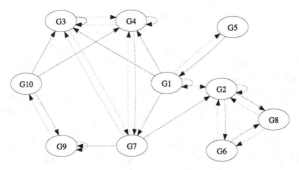

Fig. 5.11 Voting's inferred network for the DREAM4 10 gene network A. The top 25% of the ranked edges were used. Solid lines are correct guesses, dotted lines are missed edges, and dashed lines are incorrect guesses. In the top 25% of edges, Voting is able to infer some relationships involving G1 and G10, which many algorithms had previously missed. It also infers a symmetric relationship between many edges, even when that symmetry is incorrect.

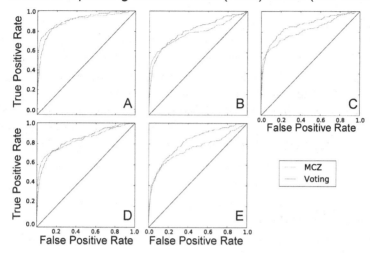

Fig. 5.12 Receiver operating characteristic curves (ROC) of the voting algorithm and median-corrected z-scores (MCZ) for the DREAM4 100 gene networks A-E. Voting was generally tied with MCZ on networks A, B, and D. On networks C and E voting surpassed MCZ. (Network C AUROC for MCZ = 0.82, voting = 0.88) (Network E AUROC for MCZ = 0.75, voting = 0.83)

5.6 Ensemble 2: Simulated Annealing

5.6.1 What it does

Simulated annealing [17] is a heuristic optimization procedure that mimics the process of annealing metal. The basic idea is that the algorithm relies on a "temperature" that is slowly reduced over time. When the temperature is high, the algorithm can make large moves with some probability. Over time, the probability of making large moves becomes smaller and smaller, and small moves become more and more likely. By "cooling" the size of the moves in this way, we can avoid getting stuck in local minima.

Rating: DREAM Good

5.6.2 The Data

The data are split up into training and test data. The training data consist of the various data types that the algorithms being used in simulated annealing require, along with gold standards that can be tested against. In our case, we will be testing using the DREAM4 data, which include gold standards for each dataset. The test data is the dataset from which we would like to infer a network. In practice, this data does not have a gold standard. However, in order to evaluate this approach we will use the gold standard for the test data, but we will not train on the test data nor its gold standard.

5.6.3 The Strategy

The basic procedure is: (i) run each of our training datasets through a list of selected algorithms that return ranked lists of edges, (ii) assign a weight to each algorithm's output and (iii) adjust the weights in order to maximize the AUROC for a consensus network built by summing the weighted ranked lists of edges. The weights are adjusted using simulated annealing.

The algorithm begins with a random set of weights. The "energy" of that set of weights is calculated according to a cost function, which in our case is $1 - \mathrm{AUROC}(w)$, where $\mathrm{AUROC}(w)$ is the AUROC of the set of weights w.

At each iteration, the energy is calculated for both the current set of weights and the neighbor. We decide whether or not to "move" to the neighbor based on an acceptance function. The acceptance function is based on the current temperature and on the difference in energy between the neighbor and current weights. If the neighbor has lower energy than the current weights (i.e., fits the data better), we move

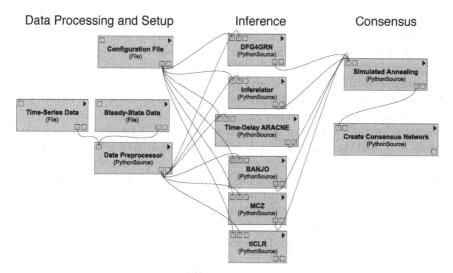

Fig. 5.13 The basic workflow diagram for the simulated annealing consensus algorithm described in Section 5.6. The inference algorithms listed could be any number and combination of algorithms that the available data permit.

to it. However, if the current weights have lower energy than the neighbor, we will move to the neighbor with some probability based on the temperature (higher temperature means more likely to move) and the difference between the energies (we are less likely to move when the neighbor is much worse than the current weights).

The idea behind sometimes moving to a worse set of weights is that we want to avoid getting stuck in a local minimum early on. When the temperature is high, we are likely to "jump out" of a local minimum. As the temperature cools, we settle into a local minimum and stay there, trying to find the best set of weights in the local area by searching immediately around the same point.

In our case, the weights correspond to coefficients on algorithms. If the weight is large for an algorithm, then that algorithm's edges receive more of a vote than for a lower-weighted algorithm. The consensus network is a sum of weighted networks. For example, suppose we run two algorithms on two training datasets. This results in four ranked lists of edges (two from each algorithm). If we group the inferred networks from each algorithm together, we have two groups N_1 and N_2. We then want to build a consensus network C from a weighted of N_1 and N_2, so $C = w_1 N_1 + w_2 N_2$. The weights can be thought of as an importance score for each algorithm.

5.6.4 Performance on examples

The DREAM4 10 and 100 gene *in silico* datasets were used to test simulated annealing. Each of the DREAM4 datasets consists of 5 networks to infer, and each network has associated knockout, knockdown, time-series, multifactorial, and wild-type data. Each network was tested by applying the weights obtained from training the simulated annealing algorithm on the other 4 datasets. The algorithms used in the test were Dynamic Factor Graphs (DFG), Gene Network Inference with Ensemble of trees (GENIE3), Median-Corrected Z-Scores (MCZ), Time-Lagged Context Likelihood of Relatedness (tlCLR), Context Likelihood of Relatedness (CLR), and Network Identification by Multiple Regression (NIR). As mentioned in Section 3.2, the idea behind choosing these algorithms was that MCZ scored the best in testing above, so it would be a baseline good answer for the annealing algorithm. The rest, being weaker learners than MCZ, would act as supplements to MCZ's result that take into account different theoretical approaches and use different types of data.

Simulated annealing was on par with MCZ for inferring the topology of the 10 gene network. As illustrated in figure 5.14, for networks A-C, the results were almost the same. Simulated annealing scored slightly higher in each case, but only by a marginal amount. In network D, median-corrected z-scores outperformed simulated annealing, and in network E, simulated annealing outperformed median-corrected z-scores. The simulated annealing algorithm performed marginally better than median-corrected z-scores when averaged across all of the networks.

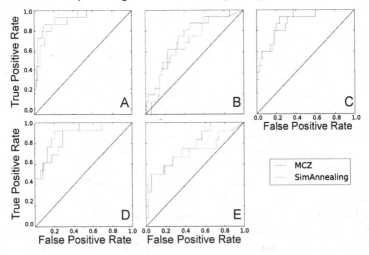

Fig. 5.14 Receiver operating characteristic curves (ROC) of simulated annealing and median-corrected z-scores (MCZ) for the DREAM4 10 gene networks A-E. Graphs A-C show results that are roughly equal, with simulated annealing performing slightly better than MCZ. Graph D shows a case where MCZ outperforms simulated annealing (AUROC for MCZ = 0.90, simulated annealing = 0.84). Graph E shows a case where simulated annealing outperforms MCZ (AUROC for MCZ = 0.71, simulated annealing = 0.80).

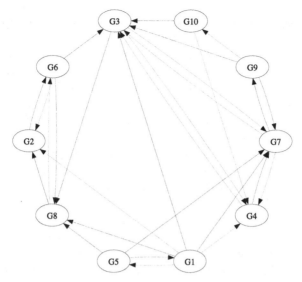

Fig. 5.15 Simulated annealing's inferred network for the DREAM4 10 gene network A. The top 25% of the ranked edges were used. Solid lines are correct guesses, dotted lines are missed edges, and dashed lines are incorrect guesses. In the top 25% of edges, Simulated annealing is able to correctly infer relationships between many genes, missing only 3.

Simulated annealing consistently outperformed MCZ on each network in the 100 gene dataset (figure 5.16), but only by a little.. MCZ's weight was the largest across all of the experiments, with the other algorithms acting as supplements to MCZ's result. This is expected. Since MCZ is generally the strongest performer, we would expect it to obtain the highest weight and use the other algorithms as supplements to help fine-tune weights. These results show that weighting different algorithms to build a consensus network amongst them works in practice, and that the "wisdom of the crowds" ([21]) can outperform any single algorithm. These results suggest that a consensus vote may be generally more reliable for larger networks, as also pointed out in [21].

While simulated annealing generally has higher ROC scores than the simple voting procedure, the difference between their scores is small. So, if a gold standard is available, simulated annealing is worthwhile. If not, then simple voting works well.

Receiver Operating Characteristic (ROC) Curve (100 Genes)

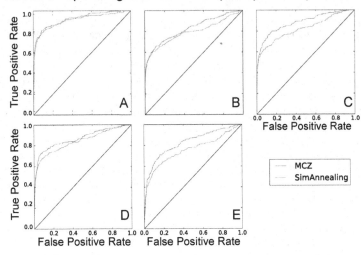

Fig. 5.16 Receiver operating characteristic curves (ROC) of simulated annealing and median-corrected z-scores (MCZ) for the DREAM4 100 gene networks A-E. Simulated annealing outperformed MCZ on each network, especially on networks C and E. (Network C AUROC for MCZ = 0.82, simulated annealing = 0.88) (Network E AUROC for MCZ = 0.75, simulated annealing = 0.83)

5.7 Conclusions

The pipeline experiments in this chapter show that combining steady state techniques with dynamic ones improve the performance of the dynamic algorithms substantially. Of the methods we report, the Inferelator pipeline performs the best, but there are many possibilities we did not test such as combining a Bayesian time-series approach using a prior network from steady-state algorithm, or integrating the topology from an ensemble algorithm with one that infers dynamics.

The ensemble experiments show that simple voting among decent algorithms performs nearly as well as or better than any individual algorithm on most data. Above all, voting reduces the likelihood of really bad networks. Voting with simulated annealing works even better, but requires the presence of gold standard data. Other techniques such as stochastic gradient descent might improve on simulated annealing. That is an area for future research.

References

1. Arthur, D., Vassilvitskii, S.: k-means++: The Advantages of Careful Seeding. Society for Industrial and Applied Mathematics (2007)
2. Bailey, T.L., Elkan, C.: Fitting a mixture model by expectation maximization to discover motifs in biopolymers. Proc Int Conf Intell Syst Mol Biol **2**, 28–36 (1994)
3. Bowers, P.M., Pellegrini, M., Thompson, M.J., Fierro, J., Yeates, T.O., Eisenberg, D.: Prolinks: a database of protein functional linkages derived from coevolution. Genome Biology **5**(5), R35 (2004)
4. Breiman, L.: Random Forests. Machine Learning **45**(1) (2001)
5. Cover, T.M., Thomas, J.A.: Elements of Information Theory. Cover/Elements of Information Theory, Second Edition. John Wiley & Sons, Inc., Hoboken, NJ, USA (2005)
6. Dempster, A., Laird, N., Rubin, D.: Maximum Likelihood from Incomplete Data via the EM Algorithm. Journal of the Royal Statistical Society. Series B (Methodological) **39**(1), 1–38 (1977)
7. Efron, B., Hastie, T., Johnstone, I., Tibshirani, R.: Least angle regression. Annals of Statistics **32**(2), 407–451 (2004)
8. Faith, J.J., Hayete, B., Thaden, J.T., Mogno, I., Wierzbowski, J., Cottarel, G., Kasif, S., Collins, J.J., Gardner, T.S.: Large-scale mapping and validation of Escherichia coli transcriptional regulation from a compendium of expression profiles. PLoS biology **5**(1), e8 (2007)
9. Gardner, T.S., Di Bernardo, D., Lorenz, D., Collins, J.J.: Inferring Genetic Networks and Identifying Compound Mode of Action via Expression Profiling. Science **301**(5629), 102–105 (2003)
10. Girolami, M.: A variational method for learning sparse and overcomplete representations. Neural computation **13**(11), 2517–2532 (2001)
11. Greenfield, A., Madar, A., Ostrer, H., Bonneau, R.: DREAM4: Combining Genetic and Dynamic Information to Identify Biological Networks and Dynamical Models. PloS one (2010)
12. Gregoretti, F., von Belcastro, Di Bernardo, D., Oliva, G.: PLoS ONE: A Parallel Implementation of the Network Identification by Multiple Regression (NIR) Algorithm to Reverse-Engineer Regulatory Gene Networks. PloS one (2010)
13. Heckerman, D., Geiger, D.: Learning Bayesian Networks The Combination of Knowledge and Statistical Data. Machine Learning (1995)
14. Hochreiter, S., Bodenhofer, U., Heusel, M., Mayr, A., Mitterecker, A., Kasim, A., Khamiakova, T., Van Sanden, S., Lin, D., Talloen, W., Bijnens, L., Gohlmann, H.W.H., Shkedy, Z., Clevert, D.A.: FABIA: factor analysis for bicluster acquisition. Bioinformatics **26**(12), 1520–1527 (2010)
15. Huynh-Thu, V.A., Irrthum, A., Wehenkel, L., Geurts, P.: Inferring Regulatory Networks from Expression Data Using Tree-Based Methods. PloS one **5**(9), e12,776 (2010)
16. Kanehisa, M., Goto, S.: KEGG: kyoto encyclopedia of genes and genomes. Nucleic Acids Research **28**(1), 27–30 (2000)
17. Kirkpatrick, S., Gelatt, C.D., Vecchi, M.P.: Optimization by simulated annealing. Science **220**(4598), 671–680 (1983)
18. Koller, D., Friedman, N.: Probabilistic Graphical Models: Principles and Techniques. MIT Press (2009)
19. Krouk, G., Mirowski, P., LeCun, Y., Shasha, D.E., Coruzzi, G.M.: Predictive network modeling of the high-resolution dynamic plant transcriptome in response to nitrate. Genome Biology **11**(12), R123 (2010)
20. Madar, A., Greenfield, A., Vanden-Eijnden, E., Bonneau, R.: DREAM3: network inference using dynamic context likelihood of relatedness and the inferelator. PloS one **5**(3), e9803 (2010)
21. Marbach, D., Costello, J., Kuffner, R., Prill, R., Camacho, D.M., Vega, N.M., Allison, K.R., Consortium, t.D., Kellis, M., Collins, J.J., Stolovitzky, G.: Wisdom of crowds for robust gene network inference. (2012)

J. M. Lingeman and D. Shasha, *Network Inference in Molecular Biology,*
SpringerBriefs in Electrical and Computer Engineering,
DOI: 10.1007/978-1-4614-3113-8, © The Author(s) 2012

22. Marbach, D., Schaffter, T., Mattiussi, C.: Generating realistic in silico gene networks for performance assessment of reverse engineering methods. Journal of Comp. Biology **16**(2), 229–239 (2009)
23. Mellor, J.C., Yanai, I., Clodfelter, K.H., Mintseris, J., DeLisi, C.: Predictome: a database of putative functional links between proteins. Nucleic Acids Research **30**(1), 306–309 (2002)
24. Palmer, J., Kreutz-Delgado, K., Wipf, D., Rao, B.D.: Variational EM algorithms for non-Gaussian latent variable models. In: Advances in Neural Information Processing Systems 18, pp. 1059–1066 (2006)
25. Pearl, J.: Causal inference in statistics: An overview. Statistics Surveys **3**, 96–146 (2009)
26. Reiss, D.J., Baliga, N.S., Bonneau, R.: Integrated biclustering of heterogeneous genome-wide datasets for the inference of global regulatory networks. BMC Bioinformatics **7**(1), 280 (2006)
27. Schaffter, T., Marbach, D., Floreano, D.: GeneNetWeaver: in silico benchmark generation and performance profiling of network inference methods. Bioinformatics **27**(16), 2263–2270 (2011)
28. Schwarz, G.: Estimating the Dimension of a Model. The annals of statistics **6**(2), 461–464 (1978)
29. Tibshirani, R.: Regression shrinkage and selection via the lasso. Journal of the Royal Statistical Society. Series B (Methodological) **58**(1), 267–288 (1996)
30. Yu, J.: Advances to Bayesian network inference for generating causal networks from observational biological data. Bioinformatics **20**(18), 3594–3603 (2004)
31. Zavlanos, M.M., Julius, A.A., Boyd, S.P., Pappas, G.J.: Inferring Stable Genetic Networks from Steady-State Data. Automatica **47**, 1113–1122 (2011)
32. Zoppoli, P., Morganella, S., Ceccarelli, M.: TimeDelay-ARACNE: Reverse engineering of gene networks from time-course data by an information theoretic approach. BMC Bioinformatics (2010)

Index

J. M. Lingeman and D. Shasha, *Network Inference in Molecular Biology*,   99
SpringerBriefs in Electrical and Computer Engineering,
DOI: 10.1007/978-1-4614-3113-8, © The Author(s) 2012